KEY PERFORMANCE INDICATORS FOR FEDERAL FACILITIES PORTFOLIOS

Authored by John H. Cable and Jocelyn S. Davis in conjunction with the Federal Facilities Council Ad Hoc Committee on Performance Indicators for Federal Real Property Asset Management

Federal Facilities Council Technical Report #147

THE NATIONAL ACADEMIES PRESS
Washington, D.C.
www.nap.edu

NOTICE

The Federal Facilities Council (FFC) is a continuing activity of the Board on Infrastructure and the Constructed Environment of the National Research Council (NRC). The purpose of the FFC is to promote continuing cooperation among the sponsoring Federal agencies and between the agencies and other elements of the building community in order to advance building science and technology—particularly with regard to the design, construction, acquisition, evaluation, and operation of Federal facilities. The sponsor agencies are the:

- Architect of the Capitol
- Department of Agriculture, Agricultural Research Service
- Department of the Air Force, Air National Guard
- Department of the Air Force, Office of the Civil Engineer
- Department of the Army, Army Corps of Engineers
- Department of the Army, Assistant Chief of Staff for Installation Management
- Department of Commerce, Office of Real Estate
- Department of Defense, Defense Facilities Directorate
- Department of Energy, National Nuclear Security Administration
- Department of Energy, Office of Engineering and Construction Management
- Department of Energy, Office of Science
- Department of Health and Human Services, Indian Health Service
- Department of Health and Human Services, National Institutes of Health
- Department of Homeland Security, Customs and Border Protection
- Department of Homeland Security, Federal Emergency Management Administration
- Department of Homeland Security, U.S. Coast Guard
- Department of the Interior, Office of Managing Risk and Public Safety
- Department of Justice, Facilities and Administrative Services
- Department of the Navy, Naval Facilities Engineering Command
- Department of State, Bureau of Overseas Buildings Operations
- Department of Veterans Affairs, Office of Facilities Management
- Environmental Protection Agency
- General Services Administration, Public Buildings Service
- National Aeronautics and Space Administration, Facilities Engineering and Real Property Division
- National Institute of Standards and Technology, Building and Fire Research Laboratory
- National Science Foundation
- Smithsonian Institution, Facilities Engineering and Operations
- U.S. Postal Service, Engineering Division

As part of its activities, the FFC periodically publishes reports that have been prepared by committees of government employees. Because these committees are not appointed by the NRC, they do not make recommendations, and their reports are considered FFC publications rather than NRC publications.

For additional information on the FFC program and its reports, visit the Web site at *www.nationalacademies.org/ffc*; write to Director, Federal Facilities Council, 500 Fifth Street, N.W., Room 944, Washington, DC 20001; or call 202-334-3374.

Printed in the United States of America 2004

FEDERAL FACILITIES COUNCIL AD HOC COMMITTEE ON PERFORMANCE INDICATORS FOR FEDERAL REAL PROPERTY ASSET MANAGEMENT

COMMITTEE CO-CHAIRS

Eduard Dailide, P.E., Office of Engineering and Construction Management, Department of Energy
Eugene Hubbard, P.E., Facilities Engineering and Real Property Division, NASA

COMMITTEE MEMBERS

Patrick Barry, P.E., Agricultural Research Service, Department of Agriculture
Robert Carlsen, Naval Facilities Engineering Command, U.S. Navy
Anthony Clifford, Division of Engineering Services, National Institutes of Health
Joseph Corliss, Division of Facilities Planning and Construction, Indian Health Service
James Curtis, Bureau of Overseas Buildings Operations, Department of State
James Dempsey, P.E., Shore Facilities Capital Asset Management, U.S. Coast Guard
Clair Gill, Facilities Engineering and Operations, Smithsonian Institution
Michael Greenan, Office of Enterprise Asset Management, Department of Veterans Affairs
David Hammond, Shore Facilities Capital Asset Management, U.S. Coast Guard
Jay Janke, Office of Secretary of Defense, Department of Defense
Michael Kastle, P.E., Office of Managing Risk and Public Safety, Department of the Interior
Barbara Nichols, Facilities and Administrative Services, Department of Justice
Scott Robinson, P.E., Facilities Engineering and Real Property Division, NASA
Dennis Sheils, Office of Facilities Management, Department of Veterans Affairs
Harry Singh, Naval Facilities Engineering Command, U.S. Navy
Edmund Tupay, P.E., Office of Real Property Asset Management, Department of Homeland Security
Louis Welker, P.E., Agricultural Research Service, Department of Agriculture
Tracy Wilson, Headquarters, U.S. Army Corps of Engineers
Ray Wynter, Office of Governmentwide Policy, General Services Administration

FEDERAL FACILITIES COUNCIL STAFF

Lynda Stanley, Director

FEDERAL FACILITIES COUNCIL

Lt. Gen. Henry J. Hatch, P.E., U.S. Army Corps of Engineers (Retired), Chair
William W. Brubaker, Facilities Engineering and Operations, Smithsonian Institution, Vice Chair
Patrick Barry, P.E., Agricultural Research Service, Department of Agriculture
Tony Clifford, Division of Engineering Services, National Institutes of Health
Will Colston, Bureau of Overseas Buildings Operations, Department of State
Captain José Cuzme, P.E., Division of Facilities Planning and Construction, Indian Health Service
Jesus de la Garza, Ph.D., Directorate for Engineering, National Science Foundation
David Eakin, P.E., Office of the Chief Architect, Public Buildings Service, General Services Administration
Ramon Garcia, Facilities Engineering Division, Customs and Border Protection, Department of Homeland Security
James Hill, Ph.D., Building and Fire Research Laboratory, National Institute of Standards and Technology
Eugene Hubbard, P.E., Facilities Engineering and Real Property Division, National Aeronautics and Space Administration
Michael Kastle, P.E., Office of Managing Risk and Public Safety, U.S. Department of the Interior
Ben Lawless, Facilities Division, Air National Guard Readiness Center
Raymond Lynn, Headquarters, U.S. Army Corps of Engineers
Capt. Jay Manik, Commandant, U.S. Coast Guard
Robert L. Neary, Jr., Office of Facilities Management, Department of Veterans Affairs
John Nerger, Facilities and Housing Directorate, Assistant Chief of Staff for Installation Management, U.S. Department of the Army
Ralph Newton, Defense Facilities Directorate, Department of Defense
Dale Olson, Office of the Civil Engineer, U.S. Air Force
Wade Raines, Engineering Division, U.S. Postal Service
James Rispoli, Office of Engineering and Construction Management, Department of Energy
Bruce Scott, National Nuclear Security Administration, Department of Energy
Stan Walker, Shore Facilities Capital Asset Management, U.S. Coast Guard
Jim Woods, Office of Real Estate, Department of Commerce
James Wright, Chief Engineer, Naval Facilities Engineering Command, U.S. Navy
John Yates, Office of Science, Department of Energy

Contents

Executive Summary   1

1. **Introduction**   5
    Background, 5
    Performance Measurement, 6
    Problem Statement and Study Objectives, 8
    Study Approach, 8

2. **Facilities Asset Management and Performance Goals**   10
    Performance Goals, 11
    Developing Performance Indicators for Facilities Portfolios, 12
    Findings, 14

3. **Existing Performance Indicators for Federal Facilities Portfolios**   16
    What Facilities Do We Have?, 16
    What Condition Are They In?, 17
    What Facilities Are Needed to Support the Organization's Missions? What Problems
      and Issues Need to Be Addressed?, 18
    How Much Are We Investing? How Much Do We Need to Invest?, 19
    What Are the Results or Outcomes of Those Investments? What Are the
      Outcomes of Decisions Not to Invest?, 21
    Findings, 25

4. **Additional Performance Indicators for Federal Facilities Portfolios**   27
    Performance Measurement Models, 27
    Additional Performance Indicators for Consideration, 29
    Findings, 30

References 32

Appendixes
A Executive Order Federal Real Property Asset Management 35
B Biographies of Consultants 40
C Engineered Management Systems and BCI 42
D Space Utilization Index 44

Acronyms

ACI	Asset Condition Index
APPA	Association of Higher Education Facilities Officers (formerly Association of Physical Plant Administrators)
AUI	Asset Utilization Index
BCI	Building Condition Index
BMAR	Backlog of Maintenance and Repair
CRV	current replacement value
DoD	Department of Defense
DOE	Department of Energy
DM	Deferred Maintenance
EMS	Engineered Management System
ERDC-CERL	Engineering Research and Development Center-Construction Engineering Research Laboratory
FCI	Facilities Condition Index
FFC	Federal Facilities Council
FRR	Facilities Revitalization Rate
FSM	Facilities Sustainment Model
GAO	Government Accountability Office, formerly General Accounting Office
GPRA	Government Performance and Results Act of 1993
GSA	General Services Administration
GSF	gross square feet

IRR	Installations Readiness Report
MDI	Mission Dependency Index
NASA	National Aeronautics and Space Administration
NNSA	National Nuclear Security Administration
NRC	National Research Council
OMB	Office of Management and Budget
PMA	President's Management Agenda
PRV	Plant Replacement Value
RR	Recapitalization Rate
SAM	Strategic Assessment Model
SR	Sustainment Rate
SUI	Space Utilization Index

Executive Summary

BACKGROUND

Much has been written about the establishment and use of performance measurement systems. Ultimately, an effective performance measurement system should support informed decision making about the allocation of resources within and by an organization. Key components of an effective system include

- Clearly defined, actionable, and measurable goals that cascade from organizational mission to management and program levels to individual performance;
- Cascading key performance indicators that can be used to measure how well mission, management, program, and individual goals are being met;
- Established baselines from which progress toward attainment of goals can be measured;
- Accurate, repeatable, and verifiable data; and
- Feedback systems to support continuous improvement of an organization's processes, practices, and results (outcomes).

Over the last 10-15 years, facilities management in both the private and public sectors has been evolving from a discipline historically focused on individual buildings to one focused on the total performance of an inventory of buildings (or portfolio) in support of an organization's overall mission. This evolving discipline is often referred to as facilities asset management.

In September 2002 the Federal Facilities Council of the National Research Council authorized a study to identify key performance indicators that could be used by senior-level federal managers to determine a full range of financial and nonfinancial results (outcomes) of investments in portfolios of facilities and to improve facilities asset management.

To make informed decisions about facilities investments and management of large inventories of facilities, senior federal executives require information that will allow them to answer such questions as

- What facilities do we have?
- What condition are they in?
- What facilities are needed to support the organization's missions?

- What problems and issues need to be addressed?
- How much are we investing? How much do we need to invest?
- What are the results or outcomes of those investments? What are the outcomes of decisions not to invest?

Subsequent to the start of the study, the U.S. General Accounting Office (recently renamed the Government Accountability Office) designated the management of federal real property as a government-wide high risk area. And on February 4, 2004, President Bush signed Executive Order 13327, Federal Real Property Asset Management, which is intended "to promote the efficient and economical use of America's real property assets and to assure management accountability for implementing Federal real property management reforms." The Executive Order specifically calls for the establishment of

> appropriate performance measures to determine the effectiveness of Federal real property management. Such performance measures shall include, but are not limited to, evaluating the costs and benefits involved with acquiring, repairing, maintaining, operating, managing, and disposing of Federal real properties at particular agencies. . . . The performance measures shall be designed to enable the heads of executive branch agencies to track progress in the achievement of Government-wide property management objectives, as well as allow for comparing the performance of executive branch agencies against industry and other public sector agencies.

Concurrent with the issuance of the Executive Order, a new program initiative for federal real property asset management was added to the President's Management Agenda.

STUDY APPROACH

The Federal Facilities Council established the Ad Hoc Committee on Performance Indicators for Federal Real Property Asset Management to provide direction and oversight for the study and to collaborate with other federal personnel and staff. Beginning in May 2003, the Ad Hoc Committee refined the study scope of work and gathered data on facilities portfolio-level performance indicators in use or under development. The consulting team of John H. Cable and Jocelyn S. Davis of Nelson Hart LLC, a team experienced in the development of performance indicators, was hired to work with the Ad Hoc Committee and author this report.

Ten meetings and work sessions were held over the course of the study. The senior representatives of the Federal Facilities Council as well as the members of the Ad Hoc Committee reviewed the final draft of the report.

FINDINGS

Finding 1: To improve decision-making about facilities investments and to improve management of federal facilities portfolios, it is important that agencies track (1) performance measures that characterize their facilities portfolios; (2) the level of alignment of their portfolios with their organizational missions; (3) investment levels; and (4) the results or outcomes of their investments.

Federal departments and agencies are at different levels of sophistication and development with respect to performance measurement systems for facilities asset management. The variation in facilities asset management systems is not surprising given the wide variation in the roles, missions, and facilities portfolios of federal agencies.

Most agencies maintain a centralized database with information about the number, type, location, age, size (typically in square feet or other appropriate units of measure), and value of their existing facilities, measures that characterize their facilities portfolios. However, the accuracy, integrity, and completeness of the information within existing databases vary. If departments and agencies are to develop effective performance measurement systems, accurate and complete data for these types of facilities portfolio characteristics are required. Efforts are already underway within the federal government to address this issue.

Finding 2: A first step in developing high-level, portfolio-oriented performance indicators is to establish organizational goals in support of mission requirements and to establish a time period for attainment.

Goals for facilities asset management should be tied to the attainment of organizational goals. Organizational goals should cascade to strategic goals, to functional unit goals, to team goals, and to individual performance goals. Several of the agencies participating in this study, including the Department of Energy's National Nuclear Security Administration, the Department of Defense, and the Department of Veterans Affairs, have established goals for portfolio facilities management and a time frame for attaining them. Those agencies have also developed performance indicators to measure progress in meeting their goals.

Finding 3: Investments made in portfolios of facilities are not often immediately visible or measurable but are manifest over a period of years. To understand the results or outcomes of facilities investments, a set of performance indicators should be tracked over a period of years and be compared to a baseline to determine whether the situation is improving or deteriorating.

Because of the long-term nature of facilities and facilities investments, snapshot reporting (where performance is now) is insufficient to understand whether facilities investments and management changes are resulting in desired outcomes. Trend reporting, reflecting historical performance in relation to organizational goals, is essential.

Finding 4: The operating environment within which a performance measurement system is used affects the types of indicators developed and their utility. Operating measures that are routine in the corporate facilities environment may not reflect the differing missions of public- and private-sector organizations and may require data not currently captured in federal accounting and management systems.

Private-sector organizations, in general, have an overall organizational goal of producing a profit. They have flexibility to design their financial systems to gather the types of data needed to track and evaluate facilities investments, operations, and management.

In the federal government the overall goal is to deliver goods and services to the public; making a profit is typically not an objective. All federal departments and agencies are subject to the same budget procedures, and their accounting systems are typically designed to track appropriations and expenditures for broad programmatic categories, not for specific assets like facilities. Other factors also come into play. For example, the lack of metering on many federal buildings inhibits tracking of utility costs, a component of operating costs. Thus, for a variety of reasons, care should be taken in developing key performance indicators for federal facilities by reference to the private sector.

Finding 5: The General Services Administration, whose mission, funding sources, and facilities portfolio are unique among government agencies, has developed performance indicators that are similar to those used by private-sector organizations.

The GSA functions, in part, as a landlord to other federal agencies. Its portfolio of facilities primarily includes office buildings and courthouses located on individual sites in hundreds of municipalities, which distinguishes it from other agencies. These factors need to be taken into account if measures used by the GSA are considered for use in other agencies. GSA's measures include Cost per Square Foot (owned); Cost per Square Foot (leased), Employees Housed; Cost Per Person; Customer Satisfaction; Vacancy Rate; Non Revenue Producing Space; Net Income; and Funds from Operations. In the GSA, trends in space demand by tenant agencies, are measured historically in square feet, and anticipated requirements are estimated in the same manner.

Finding 6. A variety of facilities portfolio-level performance indicators are being used by individual agencies to measure various aspects of facilities asset management.

These performance indicators include a Facilities Condition Index; Asset Utilization Index (an indicator used to measure the portfolio against mission requirements); Current Replacement Value (an indicator of the total amount of money invested in the portfolio); Plant Replacement Value (the cost to replace facilities assets using today's construction costs and building standards); and Sustainment Rate (a measure of the adequacy of funding for maintenance and repair); among others.

Finding 7: The performance indicator used by the greatest numbers of agencies is the Facility Condition Index (FCI), also called Asset Condition Index. Various approaches are used to calculate the FCI and to report condition-related information.

The FCI is a method for measuring the current condition of facilities to assess how much work, if any, is recommended to maintain or change the condition to acceptable levels to support missions. The calculation of FCI varies by agency. What constitutes an acceptable level of condition also varies by agency, by mission, by the importance of specific facilities (e.g., mission critical, mission supportive, mission neutral) and/or by types of facilities. Agencies also use a range of techniques to convey FCI-related information to executive management.

Finding 8: A base set of key performance indicators for measuring the outcomes of facilities investment and management within Federal agencies could include total number and size of facilities; general types; median age; geographic dispersal; Current Replacement Value; Plant Replacement Value; FCI or Installations Readiness Report; Deferred Maintenance; Asset Utilization Index; Sustainment Rate or NRC Guideline; Facilities Revitalization Rate; and Recapitalization Rate.

Used in combination and tracked against baselines over time, these indicators would help to measure:

- Improvement or deterioration in the overall condition of an organization's facilities portfolio;
- Increases or decreases in the size of its portfolio;
- Increases or decreases in the median age of the portfolio and the implications for the continuity or disruption of government operations;
- Adequacy of funding for facilities maintenance and repair, renewal, and replacement and the implications for overall long-term operating costs;
- Adequacy of funding for facilities maintenance and repair and the implications for the useful life of facilities;
- Level of alignment between an organization's missions and its facilities portfolio as evidenced by surplus, excess, or insufficient space; and
- The implications of surplus, excess, or insufficient space for future funding requirements.

Finding 9: Additional performance indicators for portfolio-level management are needed to measure desired outcomes for cost effectiveness, customer satisfaction, and process efficiencies. Several promising measures are under development in Federal agencies. Additional indicators could be adapted from other performance measurement systems to round out a comprehensive set of qualitative and quantitative performance indicators for federal facilities portfolio management, over time and as resources allow.

Efforts are underway within Federal agencies to develop a Mission Dependency Index, a Facilities Suitability Index, and a Building Condition Index. The Association of Higher Education Facilities Officers-APPA, the Project Management Institute, and other organizations have developed indices for Facilities Operating Current Replacement Value; Facilities Operating Gross Square Feet; Energy Usage; Energy Reinvestment; and Work Environment; among others. When choosing additional indicators for measuring Federal facilities portfolio investment outcomes, careful consideration and study should be given to the purpose to be served, how the data to support indicators will be gathered, the resources required (time, staff, funding) to gather data, whether existing accounting and management systems will require modification, and the costs and benefits of such modifications.

1

Introduction

BACKGROUND

Successful public managers are concerned with delivery of services to the public, finances, efficient operations, employee satisfaction, and the needs and wants of the community and its stakeholders. Good public managers have long recognized the need to set goals and standards, identify and capitalize on opportunities, detect and resolve difficulties, understand and improve upon processes, and document the results of public investments in programs and capital improvements.

Throughout the 1990s and continuing today, Congress has enacted legislation, the various presidential administrations have issued executive orders, and agencies have amended regulations, to institutionalize the establishment of goals and objectives and to develop performance measurement systems and processes in the federal government. The Government Performance and Results Act (GPRA) of 1993 (P.L. 103-62), for example, provides federal executives and program managers with an institutionalized commitment to (1) establish agency goals and objectives, including annual program goals and objectives; (2) specify how the agency is going to achieve those goals; and (3) demonstrate how agency and program performance in achieving those goals will be measured.

The intent of GPRA and related legislation[1] is to make federal departments and agencies more efficient (reduce delivery time), more cost effective, more responsive to the public, and more results driven (outcomes oriented).

Owing to the magnitude of the investment, the management of federally owned and leased facilities is receiving increased scrutiny from the Office of Management and Budget, the Government Accountability Office, and from individual departments and agencies. On a government-wide basis federally owned facilities are valued in the hundreds of billions of dollars. Upwards of $21 billion per year is spent on new facilities and the renovation of existing facilities, and billions more are spent on their operation and maintenance (NRC, 2004).

More than 30 federal departments and agencies with a wide range of missions and programs manage large inventories of facilities, also called portfolios. These portfolios range in size from a few hundred to more than a hundred thousand individual structures, buildings, and their supporting infrastructure. They are diverse in terms of facility types, mix of types, and geographic dispersal. An agency like the General Services Administration (GSA),

---

[1] Related legislation includes the Chief Financial Officers Act of 1990, the Federal Acquisition Streamlining Act of 1994, Title V, the Government Management Reform Act of 1994, and the Federal Financial Improvement Act of 1996.

whose role as a landlord to other agencies is unique, manages individual buildings (primarily offices and courthouses) located in hundreds of municipalities across the United States. The Bureau of Overseas Buildings Operations of the State Department, in contrast, manages compounds of embassy, housing, and office buildings located in 260 posts around the world. Others, for example, the National Institutes of Health, primarily operate one or two campus-like complexes, while the military services manage hundreds of city-like installations domestically and abroad.

The individual departments and agencies are responsible for the planning, acquisition, management, operation, evaluation, and disposal of facilities. The diversity of their missions and of their facilities portfolios affect how those portfolios are managed and how investments are tracked, measured, and evaluated. Once facilities are designed and constructed, the owner agencies primarily rely on annual budget appropriations for operations, maintenance, and recapitalization funding to keep them in good shape and fully supporting the missions for which they were intended.[2]

In January 2003 the U.S. General Accounting Office (recently renamed the Government Accountability Office, GAO) issued a report on Federal Real Property, in its High Risk Series that states

> Unfortunately, much of this vast and valuable asset portfolio presents significant management challenges and reflects an infrastructure based on the business model and technological environment of the 1950s. Many assets are no longer effectively aligned with, or responsive to, agencies' changing missions and are therefore no longer needed. Furthermore, many assets are in an alarming state of deterioration; agencies have estimated restoration and repair needs to be in the tens of billions of dollars. Compounding these problems are the lack of reliable governmentwide data for strategic asset management, a heavy reliance on costly leasing instead of ownership to meet new space needs, and the cost and challenge of protecting these assets against potential terrorism (GAO, 2003, p. 2).

On February 4, 2004, the President signed an executive order regarding Federal Real Property Asset Management (subsequently numbered 13327), which is intended "to promote the efficient and economical use of America's real property assets and to assure management accountability for implementing Federal real property management reforms." Among other actions, Executive Order 13327 specifically calls for the establishment of

> appropriate performance measures to determine the effectiveness of Federal real property management. Such performance measures shall include, but are not limited to, evaluating the costs and benefits involved with acquiring, repairing, maintaining, operating, managing, and disposing of Federal real properties at particular agencies. . . . The performance measures shall be designed to enable the heads of executive branch agencies to track progress in the achievement of Government-wide property management objectives, as well as allow for comparing the performance of executive branch agencies against industry and other public sector agencies.

The full text of Executive Order 13327 is contained in Appendix A.

Concurrent with issuance of the executive order, a new program initiative for Federal Real Property Asset Management was added to the President's Management Agenda (PMA). Issued in the summer of 2001 and subsequently updated, the PMA focuses on improving the measurement and performance of the federal government (available at *http://www.whitehouse.gov/omb/budget/fy2002/mgmt.pdf*).

PERFORMANCE MEASUREMENT

Much has been written about the establishment and use of performance measurement systems. Simply stated, the purpose of performance measurement is to help organizations understand how decision-making processes or practices led to success or failure and how that understanding can suggest improvements. Ultimately, an effective performance measurement system should support informed decision making about the allocation of resources within and by an organization.

---

[2]Some agencies, like the Smithsonian, also raise private-sector funds for facilities such as museums. Much of GSA's funding for operating existing buildings comes from the Federal Buildings Fund, a revolving fund.

Key components of an effective performance measurement system include

- Clearly defined, actionable, and measurable goals that cascade from organizational mission to management and program levels to individual performance;
- Cascading key performance indicators that can be used to measure how well mission, management, program, and individual goals are being met;
- Established baselines from which progress toward attainment of goals can be measured;
- Accurate, repeatable, and verifiable data; and
- Feedback systems to support continuous improvement of an organization's processes, practices, and results (outcomes).

Organizational Goals. Careful and consistent definition of organizational goals is a requirement for an effective performance measurement system. Organizational goals set standards for activity in areas that drive the attainment of strategic objectives. Organizational goals should be clearly defined, actionable, specific as to time for attainment, and reflective of the relative priority of the goals to the organization's missions.

Baselines. Baselines establish a condition or situation at a specific time.

Key Performance Indicators. Key performance indicators are metrics designed to match up with organizational goals. Writing in 1998, Paul Arveson suggests key features for performance indicators:

- **Leading Indicators:** forecast future trends inside and outside the organization;
- **Objective and Unbiased:** fact based, not subject to manipulation and can be repeated;
- **Normalized:** can be benchmarked against other organizations;
- **Statistically Reliable:** small margin of error;
- **Unobtrusive:** not disruptive of work or trust;
- **Inexpensive to Collect:** small sample sizes adequate;
- **Balanced:** qualitative/quantitative, multiple perspectives;
- **Appropriate:** measures the right things;
- **Quantifiable:** for ease of aggregation, calculation, and comparison;
- **Efficient:** can draw multiple conclusions out of dataset;
- **Comprehensive:** show all significant features of an organization's status; and
- **Discriminating:** small changes are meaningful.

To Arveson's list, the authors would add:

- **Action Oriented:** suggest next analysis or action step; motivate and direct action;
- **Understandable to Decision Makers:** understanding of performance indicators not dependent upon specialized facilities management knowledge; highly intuitive; and
- **Verifiable:** auditable.

Accurate Data. Underlying an effective performance measurement system are accurate, verifiable, and repeatable data. Lack of quality data can be a principal obstacle to choosing effective indicators or to implementing an effective performance measurement system (NRC, 1995). In some cases data are available, but their underlying accuracy and integrity may be suspect. For effective facilities asset management this becomes more problematic when data are rolled up from the individual building level to the portfolio level. Ideally, key performance indicators are supported by an integrated information technology system that collects data at the point of transaction and allows for a seamless rollup to the portfolio level.

Continuous Feedback. Performance measures are of limited value unless they are used in conjunction with formal and continuous feedback, or evaluation, processes. Evaluations have been defined as the systematic assessment of the operation and/or the outcomes of a program or policy, compared with explicit or implicit standards, as a means of contributing to the improvement of the program or policy (Weiss, 1998; NRC, 2004, p. 68).

PROBLEM STATEMENT AND STUDY OBJECTIVES

The senior facilities program manager for each Federal department and agency is responsible for the following key management activities relative to the department's or agency's portfolio of facilities:

- Physical control through establishment and maintenance of detailed inventories of assets: physical descriptions, quantities, locations, value, and use (**know what you have**);
- Maintenance and management of these facilities to support the achievement of the department's or agency's missions and the delivery of public goods and services (**stewardship**);
- Appropriate action to acquire, recycle, or remodel needed properties to provide suitable facilities to meet the existing and planned missions of the department or agency over an established planning horizon (**support mission requirements**);
- Prudent financial decisions relative to initial and ongoing control over the assets and maintenance of the value of properties in use or held as surplus (**make good financial decisions**); and
- Appropriate action to retire, recycle, reassign, or dispose of excess or obsolete properties as required to support mission requirements over various planning horizons (**dispose of excess or obsolete facilities**).

Federal facilities investment decisions involve multiple stakeholders, decision makers, and operating groups, including senior executives, such as department and agency heads, senior facilities program managers, budget analysts, and field engineers. The senior facilities program manager in an agency must advise the agency's senior executives on levels of investment required for facilities. He or she must also direct the development and operation of facilities portfolios and their related services within the budget allocated to them.

At the senior executive level of agency management, facilities-related decisions revolve around the allocation of resources (staff, funding, time) for portfolios of facilities: acquisition, renovation, operation, repair, and disposition of facilities. To make informed decisions, senior executives require information that will allow them to answer such questions as:

- What facilities do we have?
- What condition are they in?
- What facilities are needed to support the organization's missions?
- What problems and issues need to be addressed?
- How much are we investing? How much do we need to invest?
- What are the results or outcomes of those investments? What are the outcomes of decisions not to invest?

The objective of this study is to identify real property portfolio-level performance indicators that can be used by Federal executives to answer such questions and to help fulfill the requirements of Executive Order 13327. These same indicators should help senior facilities program managers to fulfill their facilities asset management responsibilities with confidence.

STUDY APPROACH

In September 2002 the Federal Facilities Council (FFC) of the National Research Council (NRC)[3] authorized a study to identify a set of key performance indicators that could be used by senior executives to determine a full

---

[3]The FFC is a cooperative association of 28 federal departments and agencies operating under the aegis of the National Research Council. The NRC is the operating arm of the National Academy of Sciences and the National Academy of Engineering. The FFC's mission is to identify and advance technologies, processes, and management practices that improve the performance of federal facilities over their entire life cycle, from planning to disposal.

range of financial and nonfinancial results (outcomes) of investments in portfolios of facilities. Further, the performance indicators identified should lend themselves to identifying the relationship between a given level of investment today and expected outcomes or future effects on cost avoidance, reliability, operating costs, life-cycle costs, facilities condition, space utilization, customer satisfaction, agency effectiveness, and the like.

As a first step, the FFC established the Ad Hoc Committee on Performance Indicators for Federal Real Property Asset Management to provide direction and oversight for the study and to collaborate with other federal personnel and FFC staff. Beginning in May 2003, the Ad Hoc Committee refined the study scope of work and gathered data on facilities portfolio-level performance indicators in use or under development. The consulting team of John H. Cable and Jocelyn S. Davis of Nelson Hart LLC, a team experienced in the development of performance indicators, was hired to work with the Ad Hoc Committee and author this report (detailed biographies are in Appendix B).

The Ad Hoc Committee identified points of contact in various agencies, several of whom were subsequently interviewed by the consultants. The consultants reviewed available descriptions of agencies' current and planned facilities management information systems and management indicators, and sought to find areas of commonality across agencies. Additionally, the consultants conducted several informal working sessions of the Ad Hoc Committee and attended a special briefing about ongoing work by the U.S. Coast Guard. The working sessions considered and refined the following topics: common performance indicators, characteristics of performance indicators, and a framework for identifying key performance indicators to support decision making related to investments in federal facilities.

2

Facilities Asset Management and Performance Goals

In part because of the President's Management Agenda, which is intended to move the federal government toward integrating business management principles, the area of federal facilities management is in flux. Decades of tradition are being challenged as Federal departments and agencies look for ways to tie all asset allocations directly to mission requirements.

Over the last 10-15 years, facilities management in both the private and public sectors has been evolving from a discipline historically focused on individual buildings to one focused on the total performance of a portfolio of buildings in support of an organization's overall mission. This evolving discipline is often referred to as facilities asset management, which

> helps to ensure that an organization's portfolio is aligned with its mission and public demand for services. Required elements include accurate data about the facilities portfolio; models for predicting the future requirements for and condition of these facilities and the performance attainable from them; engineering and economic decision support tools for trade-off analyses among competing investment alternatives; performance measures to evaluate the impacts of different types of actions (e.g., maintenance versus rehabilitation) and the timing of investments on the overall goals for facilities provision; and short- and long-term feedback procedures (NRC, 2004, p. 43).

Facilities asset management supports the day-to-day and long-term operations of an organization in meeting its mission. Poor facilities management results in

- inadequate facilities to support functional requirements;
- excess facilities that divert available funds from direct mission support;
- cost-inefficient facilities that waste available resources;
- aging facilities that become increasingly costly to maintain and less supportive of mission; and
- unavailable or inadequate facilities to meet anticipated needs.

Effective facilities asset management, on the other hand, consistently

- supports an organization's missions;
- anticipates the organization's facilities requirements;

TABLE 2.1 Example of Cascading Goals

Goals	Example Statement
Organizational	Operate cost effectively.
Strategic	Operating costs shall be within 5 percent of comparable industry, nongovernmental, or governmental peer group results not later than the end of fiscal year 2006.
Overall Facilities Portfolio	Cost of facilities shall be within 5 percent of comparable industry, nongovernmental, or governmental peer group results not later than the end of fiscal year 2006.
Facilities—Acquisition	Cost of facilities acquired through lease, purchase, or construction shall be within 5 percent of comparable industry, nongovernmental, or governmental peer group results not later than the end of fiscal year 2006.
Facilities—Utilities	Cost of utilities for base housing shall be within 5 percent per square foot of comparable industry, nongovernmental, or governmental peer group results not later than the end of fiscal year 2006.

- continuously assesses and adjusts the portfolio holdings to match facilities requirements in the near term and in the future;
- operates facilities cost effectively;
- predicts with reasonable accuracy the future consequences of current management decisions; and
- reports this highly specialized function in a concise and easily understandable manner to all nonfacilities managers involved in decision making.

PERFORMANCE GOALS

A first step in developing high-level, portfolio-oriented performance indicators is to assess which organizational goals are to be attained and to establish a time frame for attainment. In other words, goals for facilities asset management should be tied to the attainment of organizational goals. Organizational goals should cascade to strategic goals to functional unit goals, to team goals, and finally to individual performance goals.[1] As goals cascade through the organization they become increasingly more specific, but are entirely consistent in their support of the organizational goals. For facilities asset management in a federal organization the cascade of performance goals might work like the example shown in Table 2.1.

Careful and consistent definition of organizational goals is a requirement for an effective performance measurement system. To be effective goals must be actionable and must have a specific time frame for attainment. Examples of goal statements are illustrated in Table 2.2.

In the review of Federal agency materials related to performance measurement systems, the consultants identified several examples of actionable goal setting for facilities asset management programs. The Department of Energy's (DOE) National Nuclear Security Administration (NNSA), for instance, has set performance goals for its facilities management program as follows:

- By the end of fiscal year 2005 NNSA will stabilize its deferred maintenance.
- By the end of fiscal year 2009 NNSA will:
  —aggressively reduce deferred maintenance to within industry standards;
  —return facility conditions for mission essential facilities and infrastructure to an assessment level of good to excellent (deferred maintenance/replacement plant value less than 5 percent); and

---

[1]This concept is integral to *The Balanced Scorecard* (Kaplan & Norton, 1996), which is a structured performance measurement process that recommends establishing both financial and nonfinancial performance measures for organizations overall and for specific departments or functional groups within the organization. The Balanced Scorecard includes performance measures in four perspectives: financial, customer, internal business process, and learning and growth.

TABLE 2.2 Examples of Goal Statements

Goal Statement	Comment
Deferred maintenance for X agency shall be reduced.	This goal statement lacks an action orientation and a specific time frame for attainment.
Deferred maintenance for X agency shall be not more than 5 percent of current replacement value not later than the end of fiscal year 2006.	This goal statement is both actionable and specific as to time frame for attainment.
Deferred maintenance for X agency shall be not more than 3 percent of current replacement value for aircraft maintenance facilities and not more than 7 percent of current replacement value for base housing facilities by the end of fiscal year 2006.	This goal statement is actionable, specific as to time frame for attainment, and sets performance goals for the same indicator based on the priority of specifically defined facilities.

—have institutionalized responsible and accountable facility management processes, including budgetary ones, so that the condition of NNSA facilities and infrastructure is maintained equal to or better than industry standards.

Similarly, the Department of Defense (DoD) has established a goal of achieving full sustainment and full recapitalization levels by fiscal year 2008 and having all facilities at C-2 readiness level, on average, by the end of fiscal year 2010.

The Department of Veterans Affairs (VA) has identified departmental portfolio goals and measures as part of its Capital Asset Management System (see Table 2.3).

Not all departments and agencies reported setting actionable goals for facilities asset management. Some departments and agencies are hampered by a lack of organizational goals in establishing the derivative facilities asset management goals. This can be remedied on two fronts: (1) establishing organizational goals at the department or agency level and (2) establishing preliminary or working facilities asset management goals by reference to industry or other standards.

Once actionable goals are established, they should be periodically revised to respond to changing priorities and conditions as well as actual changes in the strategy of the organization. Ideally, such a review will occur in conjunction with a periodic review and revision of the organization's strategic plan. The authors believe that performance goals for facilities asset management should be periodically reviewed and revised as appropriate to reflect changing circumstances and organizational priorities.

DEVELOPING PERFORMANCE INDICATORS FOR FACILITIES PORTFOLIOS

It seems intuitively clear that run-down, poorly performing facilities will detract from mission effectiveness and will be increasingly expensive to operate and maintain. However, the analytics are simply not yet available to assess with reasonable accuracy how run-down is "too run-down," optimum reinvestment rates, and optimum timing for such investments.

Developing portfolio-level performance indicators for facilities is challenging in other ways. Such indicators should be meaningful and not lose impact in the aggregation process. They should concisely report performance against established goals and provide a clear focus for further analysis and action. They should also provide the basis for long-term assessment of the efficacy of current decisions. Aggregate indicators should be available to enable executives to make informed decisions about how to most efficiently support the requirements for providing space at a field command or operational level.

The operating environment within which performance measures are applied determines the types of measures developed and their utility. Private-sector organizations, in general, have an organizational goal of making a profit.

TABLE 2.3 VA Portfolio Goals & Measures—Business View

	Leases	Agreements	Buildings & Land	Equipment	IT
DECREASE OPERATIONAL COSTS					
Decrease by $ operating costs to commercial benchmarking	◉	◉	◉	◉	◉
Decrease by % assets exceeding useful economic life			◉	◉	◉
Decrease by % total cost of ownership	◉	◉	◉	◉	◉
Decrease operational costs by 1% based on inflation rate	◉		◉	◉	◉
DECREASE ENERGY UTILIZATION					
Increase renewable energy usage by % to total energy utilization			◉	◉	
Decrease by % total energy consumption (volume)			◉	◉	
Decrease by % energy unit costs			◉	◉	
DECREASE UNDERUTILIZED CAPACITY					
Decrease by % underutilized assets	◉	◉	◉	◉	◉
Increase asset sales by % and return funds to meet VA service delivery needs		◉	◉	◉	
Decrease % of vacant space	◉		◉		
INCREASE INTRA/INTER-AGENCY AND COMMUNITY BASED SHARING					
Increase % of assets shared across VA business lines	◉	◉	◉	◉	◉
Increase % of assets shared w/DOD, other federal agencies, state & local communities	◉	◉	◉	◉	◉
INCREASE REVENUE OPPORTUNITIES					
Increase revenues by % and use funds to meet service delivery needs		◉			
Decrease % of vacant space due to out leasing or shared		◉	◉	◉	◉
SAFEGUARD ASSETS					
Decrease by % designated high-risk assets	◉	◉	◉		◉
Increase compliance by % to safety, security, accessibility & accreditation	◉	◉	◉	◉	◉
MAXIMIZE HIGHEST AND BEST USE					
Increase # agreements for asset exchanges/sales to acquire replacement property better suited to mission purposes		◉	◉		
Increase # of agreements to ensure full utilization and optimum performance of assets		◉	◉		◉
Balance spending distribution by % to ensure portfolio management, the leveraging of investments or combination of investments	◉	◉	◉	◉	◉
Increase by % In Kind Consideration		◉			
Increase by % Cost avoidance/savings		◉			

They have the flexibility to design their financial systems to gather the types of data needed to track and evaluate facilities investments, operations, and management. Such data include operating and utility costs.

Best-practice private-sector organizations have long used such measures as internal rate of return, growth or decline in earnings per share, percentage of market share, and the like to measure performance in relation to mission and the desired results. They also use such operational measures as the level of customer satisfaction and the introduction of innovative products, techniques, or technologies (NRC, 2004). All these measures derive from an operating environment in which achieving a profit is paramount.

In the Federal government the overall goal is to deliver goods and services to the public; making a profit typically is not an objective. All Federal departments and agencies are subject to the same budget procedures. Accounting systems are typically designed to track appropriations and expenditures for broad programmatic categories, but not for specific facilities assets. The GAO has reported that

> Various material weaknesses related to financial systems, fundamental recordkeeping and financial reporting, and incomplete documentation continued to: (1) hamper the government's ability to accurately report a significant portion of its assets, liabilities, and costs; (2) affect the government's ability to accurately measure the full costs and financial performance of certain programs and effectively manage related operations; and (3) significantly impair the government's ability to adequately safeguard certain significant assets and properly record various transactions (GAO, 2003, p. 27).

The GAO also reports that the government's worldwide inventory of property, the only central source of descriptive data on the makeup of the real property inventory, does "not contain certain key data—such as data related to space utilization, facility condition, historical condition, historical significance, security, and age—that would be useful for budgeting and strategic management purposes" (GAO, 2003, p. 26).

Other factors also come into play. For example, the lack of metering on many federal buildings inhibits tracking of utility costs.

For all of these reasons, care must be taken in developing key performance indicators for federal facilities by reference to the private sector. Clearly some federal facilities and their functions are inherently governmental and will not find a comparable private-sector function.

Of all the federal agencies managing facilities portfolios, the GSA's mission most closely reflects that of a private-sector organization. As a landlord for other agencies, GSA uses performance indicators that are comparable to those used by private-sector organizations. Aggregate measures include **Cost per Square Foot (owned)**, **Cost per Square Foot (leased)**, and **Employees Housed** (occupying employees adjusted to estimate full-time equivalents).

The GSA's **Cost per Person Model** estimates the average cost per person in each of the following areas: real estate (space usage), telecommunications, information technology, and alternative work environment. An additional feature is a "what if" tool that calculates potential cost savings resulting from an alternative work environment, such as hoteling or desk sharing. Trends in space demand by tenant agencies, are measured historically in square feet, and anticipated requirements are estimated in the same manner. The GSA also measures **Customer Satisfaction** (based on customer surveys), **Vacancy Rate**, and **Non Revenue Producing Space**. GSA's measures include a **Net Income** calculation that focuses on its ability to produce revenue from its tenants to cover its costs of leasing or operating owned buildings. To the GSA, **Funds from Operations** is also a significant measure of management effectiveness.

Within the federal government the GSA's mission, funding sources, operating environment, and portfolio of facilities are unique. Further, its portfolio consists of a limited number of building types, which tend to be located on individual sites in hundreds of communities. These factors should be carefully considered if GSA's performance measures are suggested for use in other agencies.

The 30 other federal departments and agencies have a wide range of differing missions, differing funding sources, and more diverse portfolios of facilities, which differentiate them from the GSA and from most private-sector organizations. Considerable effort is being expended by individual agencies to develop key performance indicators that quantify the relationship between agency mission effectiveness and investments in facilities portfolios. These portfolio-level indicators are the focus of Chapter 3.

FINDINGS

(A) A first step in developing high-level, portfolio-oriented performance indicators is to establish organizational goals in support of mission requirements and to establish a time period for attainment.

Goals for facilities asset management should be tied to the attainment of organizational goals. Organizational goals should cascade to strategic goals, to functional unit goals, to team goals, and to individual performance goals. Several of the agencies participating in this study, including the Department of Energy's National Nuclear Security Administration, the Department of Defense, and the Department of Veterans Affairs, have established goals for portfolio facilities management and a time frame for attaining them. Those agencies have also developed performance indicators to measure progress in meeting their goals.

(B) The operating environment within which a performance measurement system is used affects the types of indicators developed and their utility. Operating measures that are routine in the corporate facilities environment may not reflect the differing missions of public- and private-sector organizations and may require data not currently captured in federal accounting and management systems.

Private-sector organizations, in general, have an organizational goal of producing a profit. They have flexibility to design their financial systems to gather the types of data needed to track and evaluate facilities investments, operations, and management. In the federal government the overall goal is to deliver goods and services to the public; making a profit is not typically an objective. All federal departments and agencies are subject to the same budget procedures, and their accounting systems are typically designed to track appropriations and expenditures for broad programmatic categories, not for specific assets like facilities. Other factors also come into play. For example, the lack of metering on many federal buildings inhibits tracking of utility costs, a component of operating costs. Thus, for a variety of reasons, care should be taken in developing key performance indicators for federal facilities by reference to the private sector.

(C) The General Services Administration, whose mission, funding sources, and facilities portfolio are unique among government agencies, has developed performance indicators that are similar to those used by private-sector organizations.

The GSA functions, in part, as a landlord to other federal agencies. Its portfolio of facilities primarily includes office buildings and courthouses located on individual sites in hundreds of municipalities, which distinguishes it from other federal agencies. These factors need to be taken into account if measures used by the GSA are considered for use in other agencies. GSA's measures include Cost per Square Foot (owned); Cost per Square Foot (leased); Employees Housed; Cost per Person; Customer Satisfaction; Vacancy Rate; Non Revenue Producing Space; Net Income; and Funds from Operations. In the GSA, trends in space demand, by tenant agencies, are measured historically in square feet, and anticipated requirements are estimated in the same manner.

3

Existing Performance Indicators for Federal Facilities Portfolios

As noted in Chapter 1, senior executives in Federal departments and agencies require information that will allow them to answer the following questions:

- What facilities do we have?
- What condition are they in?
- What facilities are needed to support the organization's missions?
- What problems and issues need to be addressed?
- How much are we investing? How much do we need to invest?
- What are the results or outcomes of those investments? What are the outcomes of decisions not to invest?

In the course of this study the consultants identified a number of key performance indicators being used within the various participating agencies that could help to answer these questions. These indicators are described below.

WHAT FACILITIES DO WE HAVE?

As noted in *Investments in Federal Facilities: Asset Management Strategies for the 21st Century* (NRC, 2004, pp. 34-35),

> Facilities asset management data at a minimum include inventory and attribute data. Inventory data describe elements of assets that do not change as a function of time—for example, the number, location, type, and size of facilities and the year of acquisition. Attribute data capture characteristics that do change over time, such as the demand for the facilities, usage, value, age, maintenance history (including treatment types and timing), operating and repair costs, condition, and so forth.

Many Federal departments and agencies do, in fact, have a centralized database that contains information about the number, type, location, age, size (typically in square feet or other appropriate units of measure),[1] and value of their facilities.

---

[1] The accuracy, integrity, and completeness of the information within these databases vary. If departments and agencies are to develop effective performance measurement systems, accurate and complete information on these types of facilities portfolio characteristics are required. Efforts are already underway within the Federal government to address this issue.

Number, Size, and Types of Facilities. These are indicators of the magnitude and diversity of an organization's facilities portfolio. These factors have implications for facilities asset management: Differing levels of resources and management practices are required to maintain portfolios of several hundred as opposed to several hundred thousand facilities. Portfolios with a preponderance of one type of facility, such as office buildings, may lend themselves to more standardized procedures and practices than portfolios with a wide range of facility types or a preponderance of one-of-a-kind high-tech facilities.

Location. Location-related information is indicative of the geographic concentration or dispersal of facilities, which has profound implications for the organizational structure and processes needed to manage them.

Age. The age of facilities is a key indicator of requirements for maintenance, repair, recapitalization, or replacement. As facilities age their various components and systems experience increased wear and tear and begin to break down. "The rate and onset of breakdowns increases if maintenance has been implemented haphazardly or not at all, and the operating condition deteriorates. Aging facilities require more, not less, maintenance and repair to keep them operating effectively" (NRC, 1998, p. 17).

Value. At least two differing indicators are currently being used by federal agencies to measure the value of their facilities portfolios. **Current Replacement Value (CRV)** is an indicator of the total amount of money invested in a facilities portfolio. CRV is a function of the original acquisition cost of each facility in the portfolio plus capital improvements occurring after the original construction, multiplied by an inflation factor (based, for example, on the *Engineering News-Record's* building cost index) to calculate the present value of the investment.

Plant Replacement Value (PRV) represents the cost to replace facilities assets using today's construction costs and building standards and codes. It is typically calculated as a function of the current unit construction costs (e.g., dollars per square feet) for various types of facilities, multiplied by the total number of units (e.g., square feet) of each type of facility.

WHAT CONDITION ARE THEY IN?

Federal agencies use a range of techniques and methodologies to assess the condition of their facilities. Typically, condition assessment surveys utilize trained personnel who inspect the facilities, make determinations regarding the facilities' physical condition and their performance, and identify maintenance or repair deficiencies. In most cases the results are entered into computerized maintenance management systems so that the inspection data can be reported out. Some agencies use the Engineered Management Systems (EMS) developed by the Army's Engineering Research and Development Center-Construction Engineering Research Laboratory (ERDC-CERL). The EMS is a family of tools (e.g., BUILDER, RAILER, PAVER) that aid in assessing the condition of facilities and allocating funding (see Appendix C).

Facility Condition Index (FCI) also called Asset Condition Index (ACI). To quantify the results of condition assessment surveys, many federal agencies use a Facility Condition Index performance indicator. The FCI is, in fact, the performance indicator most widely used by agencies participating in this study. The FCI is a method of measuring the current condition of facilities to assess how much work, if any, is recommended to maintain or change their condition to acceptable levels to support organizational missions. What constitutes an acceptable level of condition will vary by agency, by mission, by the importance of specific facilities (e.g., mission critical, mission supportive, mission neutral) and/or by types of facilities. This variability underlines the importance of setting performance goals for facilities asset management.

There is also variability in how the FCI is developed across departments and agencies. In the Department of Energy and the U.S. Coast Guard, FCI is calculated as Deferred Maintenance (see below) divided by Current Replacement Value. NASA's FCI uses a 5-point scale: 5 means no or few repair requirements, 1 means the facility should be or is condemned. Values are generated annually in tandem with calculation of Deferred Maintenance, using a parametric model.

Agencies also use a range of techniques to convey condition-related information to executive management: color schemes, letter grades, numerical scales, dollar scales, or ratios. These schemes often blend two different issues: the current state of facilities and users' or occupants' expectations for them. Three, four, or five color schemes are used. Some color schemes derive from a "traffic signal" concept of red-yellow-green.

Building Condition Index (BCI). The BCI has been developed as a component of the BUILDER EMS. The BCI is developed from a roll up of system component sectors to components, to systems, to a building. It could potentially be used across an entire portfolio of facilities and provide "drill down" data as well (see Appendix C), although no agency is using BCI this way today.

Deferred Maintenance (DM), also called Backlog of Maintenance and Repair (BMAR). Federal Accounting Standards Advisory Board (FASAB) Standard No. 6, as amended, requires federal agencies to annually report their total dollar amount of deferred maintenance. The FASAB standard defines deferred maintenance as "maintenance that was not performed when it should have been or was scheduled to be and which, therefore, is put off or delayed for a future period" (FASAB, 1996).

To calculate deferred maintenance some agencies systematically itemize nonroutine repair requirements at the building system level and cost these out. These costs are then aggregated at the building level, over classes or installations of buildings, and ultimately at the agency level. The agency then seeks to fund these requirements through funding accounts that have a variety of names referring to major (or minor) capital expenditures. The DOE aggregates deferred maintenance by building, by system. NASA has developed a parametric model for estimating deferred maintenance across an entire inventory of facilities.

Because agencies do not necessarily obtain sufficient funds to address all the maintenance and repair requests, a separate measure is developed called Backlogged Maintenance and Repair (BMAR). Projects that have not been funded within a year of their recognition are included in this list, and their costs in aggregate reflect a level of deferred maintenance. When they are funded they are dropped from the list. The Smithsonian Institution, and perhaps others, prioritizes BMAR into levels of urgency or overall condition for a facility.

WHAT FACILITIES ARE NEEDED TO SUPPORT THE ORGANIZATION'S MISSIONS? WHAT PROBLEMS AND ISSUES NEED TO BE ADDRESSED?

These questions are interrelated. As noted by the GAO and others, two of the major issues facing federal facilities managers today are the (1) lack of portfolio alignment as evidenced by facilities that are excess to the mission; and (2) the deteriorating condition of buildings as evidenced by the ever-growing estimates of deferred maintenance. A separate but related issue is determining the point at which it is more cost effective to replace existing facilities than to continue to operate, maintain, and repair them. Several measures have been developed to help indicate the presence of underutilized facilities, facilities excess to the mission, and those that should be replaced, as described below.

The Asset Utilization Index (AUI). Developed by the DOE, AUI is used to measure the asset inventory against mission requirements. This index is a ratio of utilized assets to total assets. Utilized assets are determined by annual surveys, and separate measures are developed for facilities versus land holdings. The AUI detects surplus space. Deficiencies in facilities quantity surface with proposed acquisitions and through the funding cycle.[2]

The BCI described above and in Appendix C also addresses the functionality of facilities and could potentially be used to help indicate facilities that are obsolete.

---

[2] The Coast Guard is developing a Space Utilization Index (SUI) that could potentially be of use to other agencies. The SUI is calculated for specific types of spaces by dividing the actual space by authorized space standards. An index of 1.15 indicates excess space; an index of less than .95 indicates insufficient space to support an activity (see Appendix D).

Vacancy Rate. Used by the GSA and the Department of Veterans Affairs, a vacancy rate measure could potentially be used more widely as an indicator of excess space. However, great care is required in calculating and interpreting a vacancy rate indicator. For example, a single aggregate number indicating that x percent of the total square footage of the facilities portfolio is vacant would not distinguish between facilities that support mission and those that might be excess. Nor would vacancy rate necessarily distinguish between short-term (turnover) vacancies and space that has been vacant for several years or longer. A single aggregate indicator also may not distinguish between concentrations of usable vacant space large enough to support an operational unit versus scattered pockets of vacant space suitable for 10 or fewer people.

The Installations Readiness Report (IRR). The IRR is used in the Department of Defense. Base Commanders provide readiness ratings for all facilities under their command, where facilities are classified into one of nine different categories. Ratings range from C-1 (ready) to C-4 (cannot support mission). The facility readiness ratings system is augmented by an IRR that flags serious deficiencies in facilities. A rating of C-4 indicates facilities that are excess to the mission or require replacement.[3] The Navy's version of the IRR also shows where the requirement is exceeded.

HOW MUCH ARE WE INVESTING? HOW MUCH DO WE NEED TO INVEST?

A senior executive in a private-sector organization is likely to ask, "What are the operating costs for our facilities?" when making investment decisions. A key performance indicator for facilities portfolios, then, is operating costs (e.g., utilities, custodial services).

In the federal government, tracking utility costs is difficult because many buildings are not metered. Tracking operating costs in general is difficult because existing budgeting and accounting systems typically are structured to track other information, such as total funds appropriated for operations, of which facilities is only one of many components. Similarly, it is difficult to determine how much money is being invested in facilities because all of the components—new construction, operations, alterations, maintenance, repairs, and demolition—are tracked through a variety of accounts and may not be easily identifiable.[4] To track such costs existing systems will likely need to be modified, an undertaking that can be resource intensive. In gathering such data or revamping accounting systems it is important to consider the trade-offs "between the amount of data collected, the frequency at which it is collected, the quality of the data, and the cost of the entire process, including data entry and storage" (Sanford and McNeil, 1997; NRC, 1998).

Lacking data for operating costs, the questions posed by senior federal executives today more often become, "How much are we investing? How much do we need to invest?" Several differing performance indicators and performance measurement systems have been developed in response to these questions.

Sustainment Rate. The Department of Defense has developed a facilities Sustainment Rate indicator and a Recapitalization Rate indicator to track annual progress toward the established goal: C-2 readiness for facilities, on average, by the end of fiscal year 2010. (The current status of facilities readiness is also tracked using the aforementioned Installations Readiness Report). The future-years defense program (the DoD's funding plan) tracks the resources related to the performance indicators, and summary performance reports for the DoD's leadership are in place. Snapshots of the Sustainment and Recapitalization Rates are taken annually at three major points: as programmed, as budgeted, and as executed.

The Sustainment Rate is the product of sustainment funding divided by sustainment requirement (SF/SR) and measures the adequacy of funding for facilities maintenance and repair (but not the adequacy of funding for capital

---

[3] Similarly, NASA's use of FCI where 1 means condemned also could be used as an indicator of excess facilities or those requiring replacement.

[4] A few agencies, such as the DoD, have devised accounting systems that do make it relatively easy to aggregate total investments in facilities. Chapter 4 of the 1998 NRC report *Stewardship of Federal Facilities: A Proactive Strategy for Managing the Nation's Public Assets* proposes an illustrative template for tracking the total ownership costs of federal facilities.

renewal and replacement). The goal is 1.0 (i.e., 100 percent). Results of less than 100 percent indicate a failure to fund current needs. Below some minimal funding level the Sustainment Rate would indicate that funding shortfalls are potentially shortening the expected service life of facilities and degrading their performance. In this sense, Sustainment Rate is a predictive indicator of the potential future consequences of today's investment decisions.

The overall sustainment requirement is computed with a department-wide Facilities Sustainment Model (FSM). The model covers all sources of sustainment funding, including sources external to the DoD. The model also covers all types of facilities, sorted into approximately 400 common categories. The sustainment requirement for each of the 400 categories is expressed in the unit of measure for that category (e.g., per square foot, gallon, ton, mile).

Sustainment requirements are based on common commercial benchmarks for each category. The benchmarks account for normal maintenance and repair tasks, including routine (i.e., expected) replacement of major facility components, required to sustain a facility through an expected service life. Each benchmark is further defined by specific tasks and a normal schedule (e.g., replace carpeting every 10 years). Individual tasks are priced for materials and labor. The Sustainment Rate indicator is also used by the military services (Army, Navy, Air Force) and is under consideration by NASA.

Recapitalization Rate. The Recapitalization Rate is the product of the expected service life divided by the funded service life (ESL/FSL). The minimum goal is 1.0 (i.e., 100 percent). Often a shorthand Recap Rate is described simply as the funded service life (e.g., 136 years); the expected service life (e.g., 67 years) is described separately as a corporate goal. The overall recapitalization requirement is computed using a department-wide facilities recapitalization metric. The current metric covers most facilities and funding sources in the DoD and is being extended to cover others (such as family housing).

Recapitalization requirements are based on internal benchmarks for expected service life. Estimated expected service life averages 67 years for the DoD as a whole when weighted by Plant Replacement Value. The benchmarks assume full sustainment levels throughout service life. If expected service life has been shortened (because of lack of sustainment or other causes), the Recapitalization Rate needs to be temporarily accelerated accordingly. The DoD has common but combined restoration and modernization programs in each military service and agency. Modernization typically addresses service-life-based recapitalization needs caused by normal aging processes; restoration typically addresses accelerated recapitalization needs caused by low sustainment rates or unforeseen events. The Recapitalization Rate indicator is also used by the military services (Army, Navy, Air Force) and by NASA.

Facilities Revitalization Rate (FRR). NASA uses a Facility Revitalization Rate to determine major repair requirements and to track requirements and funding. The FRR is an indication of how often a facility is completely revitalized. It is calculated by dividing CRV by annual facility revitalization funding. NASA's annual revitalization funding is Construction of Facilities funding minus those projects that are "new capability" or "new footprint" construction. The FRR accounts for repairs and upgrades needed because of obsolescence, modernization, aging materials, and new requirements.

In NASA, the FSM, FCI, and FRR are used together to describe a total facility maintenance, repair, and revitalization requirement. Using these indices with its parametric model, NASA can estimate how much funding is required to raise the FCI of its facilities portfolio from a level of 4.0 to 4.5 on a scale of 1 to 5.[5]

NRC Guideline. One other indicator of the level of investment used by several agencies, including the Agricultural Research Service and the Smithsonian Institution is a guideline developed by the National Research Council in 1990, often referred to as the "2 to 4 percent of CRV" guideline. This guideline states that "an appropriate

---

[5]The Association of Higher Education Facilities Officers-APPA, has developed a Capital Reinvestment Index (CRI) that seems to serve a similar purpose as the Facilities Revitalization Rate. The CRI assesses the adequacy of capital repair and replacement activities relative to the CRV and is calculated as annual capital renewal and renovation expenditures/CRV (APPA, 2001).

budget allocation for routine M&R [maintenance and repair] for a substantial inventory of facilities will typically be in the range of 2 to 4 percent of the aggregate current replacement value of those facilities (excluding land and major associated infrastructure)" (NRC, 1990, p. ix). Agencies that are investing less than 2 per cent of CRV on an annual basis are assumed to be underinvesting in facilities. The report also states that the "specific percentage for any inventory will depend on such factors as the age of the buildings in the inventory, the type of construction (permanent, temporary), the level of use of the buildings, the structure of the maintenance organization, and the climate. However, the relationship between M&R requirements and the current replacement value of single buildings may be outside the proposed range" (NRC, 1990, p. 10).

At the Smithsonian, facility conditions are color coded and then normalized to the NRC guideline. Facilities requiring routine maintenance and repair may require funding levels of 2 percent of CRV and are coded blue. Facilities in a more deteriorated condition may require investments of 3 percent of CRV and are color coded green. Severely deteriorated facilities may require investments of 5 to 6 percent of CRV and are coded red or lavender. Physical changes then are costed by systems, summed for a facility, and compared with the color standards to determine color status. Projected or expected funding is matched to repair schedules to project when color status will change in an out year.

WHAT ARE THE RESULTS OR OUTCOMES OF THOSE INVESTMENTS? WHAT ARE THE OUTCOMES OF DECISIONS NOT TO INVEST?

For any organization it is important to understand why its decision-making processes or management practices led to success or failure and how that understanding can support improvements. "Best-practice organizations measure the results or outcomes of facility investments by establishing baselines and performance measures to constantly monitor and track all aspects of operations and their results in relation to organizational objectives" (NRC, 2004, p. 65).

The stated purpose of this study is to

> identify a set of key performance indicators that could be used by senior-level managers to determine a full range of financial and nonfinancial results (outcomes) of investments in portfolios of facilities. Further, the performance indicators identified should lend themselves to identifying the relationship between a given level of investment today and expected outcomes or future effects on cost avoidance, reliability, operating costs, life-cycle costs, facilities condition, space utilization, customer satisfaction, agency effectiveness, and the like.

To understand the results or outcomes of facilities investments, Federal departments and agencies first need to establish facilities asset management performance goals that (1) derive from or are supportive of organizational goals; (2) are actionable; and (3) have a time frame for attainment. Performance indicators designed to measure how well the goals are being met can then be tracked over several years and be compared to a baseline to determine whether the situation is improving or deteriorating: Investments made in portfolios of facilities are not often immediately visible or measurable but are manifest over a period of years. Snapshot reporting (where performance is now) is insufficient to understand whether facilities investments and management changes are resulting in desired outcomes. Trend reporting, reflecting historical performance in relation to organizational goals, is essential to support effective decision making and to determine whether progress is being made toward attaining the goals.

As summarized in Table 3.1 there is wide variability among the key performance indicators being used by departments and agencies to determine the results of facilities investments. No one department or agency uses all the identified measures. Some agencies refer to an indicator by the same name but calculate it differently, for example, FCI. In some cases the same type of indicator is referred to by different names, for example, FCI and ACI.[6] And some indicators have different names and are calculated differently but serve the same purpose. For instance, Sustainment Rate and the NRC Guideline both measure the adequacy of funding for facilities mainte-

---

[6]To facilitate communication between technical and nontechnical managers and decision makers on a government-wide basis, it may be useful for the agencies to consider calling similar measures by the same name even if the measures are calculated differently.

TABLE 3.1 Summary of Existing Performance Indicators for Facilities Portfolios

Key Measurement Questions/Goal Attainment	Key Performance Indicators (KPI)	Calculation	Purpose of KPIs/Comments
What facilities do we have?	Number of facilities Type of facilities Location Age Size (in sq. ft. or other appropriate measure) Current Replacement Value (CRV) Plant Replacement Value (PRV)	Varies	Taken together, characterize an organization's facilities portfolio and implies differing management requirements. Complete, accurate, verifiable data are fundamental to performance indicator utility.
What condition are they in?	Facilities Condition Index (FCI) or Asset Condition Index (ACI)	Varies	Measures the current conditions of facilities based on pre-established, auditable criteria. An acceptable level of condition will vary by mission, by agency, by organization, and by the importance of specific facilities. This variability underlines the importance of setting performance goals.
	Deferred Maintenance (DM) or Backlog of Maintenance and Repair (BMAR)	Varies	Measures maintenance that was not performed when it was scheduled and is delayed for a future time. Indicator of potential shortening of the useful life of facilities and likely increase of long-term maintenance and repair costs.
What facilities are needed to support the organization's missions? What problems and issues need to be addressed?	Asset Utilization Index	Utilized Assets/Total Assets	Used to measure the asset inventory against mission requirements. Detects surplus space.
	FCI and Installations Readiness Report	Varies	In addition to measuring current condition, can also help to indicate facilities that are excess to mission or that should be replaced (see text).
	DM	See above	See above.
How much are we investing? How much do we need to invest?	Sustainment Rate (SR)	Sustainment Funding/Sustainment Requirements	Measures the adequacy of funding for facilities sustainment (routine maintenance and repairs expected over the useful life of assets). Below a minimal funding level, SR indicates that funding shortfalls are potentially shortening useful asset life and likely increasing long-term costs.

	Recapitalization Rate (RR)	Expected Service Life/Funded Service Life	Measures how the level of funding received is affecting the expected service life of the asset. Ratios less than 1 indicate that the funding levels are potentially shortening the useful life of assets and likely increasing long-term costs.
	Facilities Revitalization Rate (FRR)	CRV/Annual Facility Revitalization Funding	Accounts for repairs and upgrades needed due to obsolescence, modernization, aging materials, and new requirements.
	NRC Guideline (NRC)	2-4 percent of CRV of the entire facilities portfolio	Annual investments of less than 2 percent of CRV indicate underinvestment in facilities maintenance and repair, resulting in potential shortening of useful asset life and likely increasing long-term costs.
What are the results or outcomes of those investments? What are the outcomes of decisions not to invest?	Number Type Location Age Size CRV PRV FCI SR or NRC DM RR FRR	Varies	Used in combination and tracked against baselines over time, these indicators would help to measure several aspects of facilities asset management.
	Footprint Reduction	Annual Gross Square Feet (GSF) of Excess Facilities Space Reduced (and cumulative percentage of GSF reduced)	Measures progress toward attainment of a goal to reduce overall size (in sq. ft. or other appropriate measure) of the facilities portfolio.
	Deferred Maintenance Reduction	Annual Dollar Amount of Deferred Maintenance Backlog Reduced (and cumulative percentage of estimated total deferred maintenance reduced)	Measures progress toward attainment of a goal to reduce overall level of deferred maintenance.

nance and repair. Both operate on the premise that investments below a minimum level over several years will shorten the useful life of facilities and lead to degraded performance.

Because of their differing missions, the diversity of their portfolios (large/relatively small; limited/broad range of types; centralized/dispersed), and the wide range of investment levels, it is not necessary that every agency track the same performance measures or track them the same way. The cost in time and resources to reconfigure accounting systems and databases would likely far outweigh the potential benefits. However, within agencies the performance indicators in use should be calculated consistently and tracked over time in relation to established baselines.

To improve decision making about facilities investments and to improve management of federal facilities portfolios, it is important that agencies track (1) measures that characterize their particular facilities portfolio; (2) the level of alignment of that portfolio with organizational mission; (3) investment levels; and (4) the results or outcomes of those investments.

Indicators that are useful in characterizing an organization's facilities portfolio include the total number and size (in square feet or other appropriate measure), general types (e.g., administrative, laboratory, industrial), median age, geographic dispersal, value (CRV, PRV), and condition (FCI, IRR, DM).

A portfolio that is aligned with mission could be characterized as having the right facilities in the right locations at a level of condition to support day-to-day operations cost effectively. It would not include large amounts of space excess to the mission or large numbers of facilities in such a deteriorated condition that mission achievement is hindered. Indicators that would be useful in characterizing the level of alignment with mission include surplus space (AUI) and condition.

Levels of investment are important in that facilities require funding to support routine maintenance and repair to operate efficiently and cost effectively, and to perform for the duration of their projected useful life. Even with timely maintenance and repair, components wear out and require renewal or replacement over time. Some facilities require replacement due to changing technologies or materials, obsolescence, or other factors.

The Sustainment Rate (SR) or the NRC Guideline can be used to track the adequacy of funding for maintenance and repair. The Facilities Revitalization Rate or an equivalent measure can be used to track the adequacy of funding for long-term capital repair and replacement. The Recapitalization Rate, or an equivalent measure, can be used to determine whether funding should be temporarily accelerated if expected service life has been shortened due to lack of sustainment or other causes. In federal organizations where the level of funding is inadequate to perform timely maintenance, repairs, renewal, and replacement of facilities, these measures will indicate that the useful life of facilities will likely be shortened, and that long-term operating costs will likely be higher than necessary. Thus, in a sense, they can be viewed as predictive measures.

Tracked over time and in relation to performance goals and baselines, the set of indicators described above and shown in Table 3.1 (total number and size of facilities, general types, median age, geographic dispersal, CRV, PRV, FCI or IRR, DM, AUI, SR or NRC Guideline, FRR, and RR) can be used to begin to measure the results or outcomes of decisions to invest as well as the outcomes of decisions not to invest.

Combined, these indicators can measure

- Improvement or deterioration in the overall condition of an organization's facilities portfolio;
- Increases or decreases in the size of the portfolio;
- Increases or decreases in the median age of the portfolio and the implications for the continuity or disruption of government operations;
- Adequacy of funding for facilities maintenance and repair, renewal, and replacement and the implications for overall long-term operating costs;
- Adequacy of funding for facilities maintenance and repair and the implications for the useful life of facilities;
- Level of alignment of an organization's mission and its facilities portfolio as evidenced by surplus, excess, or insufficient space; and
- The implications of surplus, excess, or insufficient space for future funding requirements.

For those organizations that have specific goals to reduce deferred maintenance or reduce the overall size of the facilities portfolio, specific measures could be developed and tracked. DOE's NNSA, for example, employs **Deferred Maintenance Reduction** and **Footprint Reduction** (reduction in excess facilities) indicators. Deferred Maintenance Reduction is the annual dollar amount of deferred maintenance backlog reduced (and cumulative percentage of the estimated total deferred maintenance backlog of $1 billion to be reduced). Footprint Reduction is the annual gross square feet (GSF) of excess facilities space reduced (and cumulative percentage of GSF reduced) to achieve a total of 3 million GSF of excess facilities space reduced by fiscal year 2009.

Over time and as resources allow, additional indicators should be developed to measure outcomes for cost effectiveness, customer satisfaction, and process efficiencies. Additional measures for consideration are the subject of Chapter 4.

FINDINGS

(A) To improve decision-making about facilities investments and to improve management of federal facilities portfolios, it is important that agencies track (1) performance measures that characterize their facilities portfolios; (2) the level of alignment of their portfolios with their organizational missions; (3) investment levels; and (4) the results or outcomes of their investments.

Federal departments and agencies are at different levels of sophistication and development with respect to performance measurement systems for facilities asset management. The variation in facilities asset management systems is not surprising given the wide variation in the roles, missions, and facilities portfolios of Federal agencies.

Most agencies maintain a centralized database with information about the number, type, location, age, size (typically in square feet or other appropriate units of measure), and value of their existing facilities, measures that characterize their facilities portfolios. However, the accuracy, integrity, and completeness of the information within existing databases vary. If departments and agencies are to develop effective performance measurement systems, accurate and complete data for these types of facilities portfolio characteristics are required. Efforts are already underway within the Federal government to address this issue.

(B) Investments made in portfolios of facilities are not often immediately visible or measurable but are manifest over a period of years. To understand the results or outcomes of facilities investments, a set of performance indicators should be tracked over a period of years and be compared to a baseline to determine whether the situation is improving or deteriorating.

Because of the long-term nature of facilities and facilities investments, snapshot reporting (where performance is now) is insufficient to understand whether facilities investments and management changes are resulting in desired outcomes. Trend reporting, reflecting historical performance in relation to organizational goals, is essential.

(C) A variety of facilities portfolio-level performance indicators are being used by individual agencies to measure various aspects of facilities asset management.

These performance indicators include a Facilities Condition Index; Asset Utilization Index (an indicator used to measure the portfolio against mission requirements); Current Replacement Value (an indicator of the total amount of money invested in the portfolio); Plant Replacement Value (the cost to replace facilities assets using today's construction costs and building standards); and Sustainment Rate (a measure of the adequacy of funding for maintenance and repair); among others.

(D) The performance indicator used by the greatest number of agencies is the Facility Condition Index (FCI), also called Asset Condition Index. Various approaches are used to calculate the FCI and to report condition-related information.

The FCI is a method for measuring the current condition of facilities to assess how much work, if any, is recommended to maintain or change the condition to acceptable levels to support missions. The calculation of FCI varies by agency. What constitutes an acceptable level of condition also varies by agency, by mission, by the importance of specific facilities (e.g., mission critical, mission supportive, mission neutral) and/or by types of facilities. Agencies also use a range of techniques to convey FCI-related information to executive management.

(E) A base set of key performance indicators for measuring the outcomes of facilities investment and management within Federal agencies could include total number and size of facilities; general types; median age; geographic dispersal; Current Replacement Value; Plant Replacement Value; FCI or Installations Readiness Report; Deferred Maintenance; Asset Utilization Index; Sustainment Rate or NRC Guideline; Facilities Revitalization Rate; and Recapitalization Rate.

Used in combination and tracked against baselines over time, these indicators would help to measure

- Improvement or deterioration in the overall condition of an organization's facilities portfolio;
- Increases or decreases in the size of its portfolio;
- Increases or decreases in the median age of the portfolio and the implications for the continuity or disruption of government operations;
- Adequacy of funding for facilities maintenance and repair, renewal, and replacement and the implications for overall long-term operating costs;
- Adequacy of funding for facilities maintenance and repair and the implications for the useful life of facilities;
- Level of alignment between an organization's missions and its facilities portfolio as evidenced by surplus, excess, or insufficient space; and
- The implications of surplus, excess, or insufficient space for future funding requirements.

4

Additional Performance Indicators for Federal Facilities Portfolios

The base set of performance measures suggested in Chapter 3 can help to improve facilities asset management activities in the federal government. However, an expanded set of indicators to be developed in the long term is desirable. The purpose of developing additional indicators is to allow agencies to measure aspects of facilities investment, management, and outcomes that are not fully captured by the proposed base set.

The authors believe that additional key performance indicators should incorporate major management objectives of the federal government to become more efficient, more cost effective, more responsive to customers, and more results driven.

Additional key performance indicators should also

- Be easy to understand by nontechnical staff and decision makers;
- Directly and substantially support critical decision making;
- Use available or easily gathered data and understandable calculations; and
- Be supportive of, not in conflict with, the suggested set of key performance indicators described in Chapter 3.

In the course of this study, performance measurement systems in use by private industry and academia were reviewed. Some promising performance indicators being developed within federal agencies were also identified. The authors believe these systems and indicators deserve consideration in any future effort to round out a comprehensive system of qualitative and quantitative performance measures for federal facilities asset management.

PERFORMANCE MEASUREMENT MODELS

Several models for developing performance measurement systems have been developed and are in use. These include the Malcolm Baldrige National Quality Program, the Balanced Scorecard, and the Strategic Assessment Model (SAM) developed by the Association of Higher Education Facilities Officers-APPA.

The Balanced Scorecard provides a well-researched and heavily utilized approach to the development of balanced performance indicators that seek to measure both quantitative and qualitative outcomes. A key aspect of the Balanced Scorecard is its focus on four separate but related perspectives of organizational performance and management: finances, internal processes, customer satisfaction, and workforce support (called innovation and learning). The Balanced Scorecard by design provides for cascading goals, objectives, and performance measures from the organizational mission to management and program levels to individual performance. Like the Baldrige program, the Balanced Scorecard can be applied to any aspect of organizational performance.

APPA's SAM is specific to facilities management. The SAM incorporates features from both the Malcolm Baldrige National Quality Program and the Balanced Scorecard and provides a consistent vocabulary for continuous improvement. It uses a five-level rating system with criteria applicable to each level of performance. Although SAM has not yet been implemented broadly in the higher education arena, it is useful because it addresses facilities management in a complex (university) facilities environment and more closely matches the funding environment of the federal government than that of private-sector organizations.

One caveat in applying the SAM directly to a federal agency relates to the geographical distribution of university facilities: Most universities are responsible for managing facilities that are concentrated in one or two campus complexes. This is very different from the geographic distribution of the facilities portfolios of most Federal agencies. As noted in Chapter 2, the geographic distribution of facilities has implications for how they are managed. However, some of the facilities indicators developed for the SAM could be adopted or adapted for federal agency use without adopting the entire model.

The four perspectives of the Balanced Scorecard as applied in SAM are described below. (For further reading on SAM see APPA, 2001.)

The financial perspective reflects the organization's performance in ensuring its financial integrity and demonstrates stewardship responsibility for capital and financial resources associated with the operation and preservation of facilities. Financial performance indicators are tracked to ensure that services are delivered efficiently and cost effectively. The financial perspective is linked to other perspectives through the relationships between costs and the results in achieving other objectives. *An example would be to understand how improving internal processes or customer satisfaction correlates with increasing or decreasing costs. Another might be to determine how financial benefits are derived from improvements in employee safety, absenteeism, and turnover* (italics added). Primary services include those for operations and maintenance, energy and utilities, and planning, design, and construction (combines Baldrige categories 4.1, 4.2, and 7.2).

The internal process perspective addresses the key aspects of improving the organization's processes for delivering primary services, including services for operations and maintenance, energy and utilities, and planning, design, and construction. Examples of processes supporting these might include handling of work orders, procurement, billing, and relationships with suppliers. Evidence should show that processes for delivering services are efficient, systematic, and focused on customer needs (combines Baldrige categories 6.1, 6.2, 6.3 and 7.4).

The innovation and learning perspective addresses key practices directed toward creating a high-performance workplace and a learning organization. In a learning organization people at all levels are continually increasing their knowledge and their capacity to produce the best practices and best possible results. This perspective considers how the organizational culture, work environment, employee support climate, and systems enable and encourage employees to contribute effectively. Work environment and systems include work and job design, compensation, employee performance management, and recognition programs. Training is analyzed to determine how well it meets ongoing needs of employees and how well it develops their leadership and knowledge-sharing skills to improve efficiency and accommodate change. There is an emphasis on measuring results relating to employee well-being, satisfaction, development, motivation, and effectiveness and work system performance (combines Baldrige categories 5.1, 5.2, 5.3, and 7.3).

Customer perspective addresses how the organization determines requirements, expectations, and preferences of customers to ensure relevance of current services and to develop new opportunities; how the organization builds relationships with customers; and how the organization measures results of customer satisfaction and performance of services. . . . Primary services would include those for operations and maintenance, energy and utilities, and planning, design, and construction (combines Baldrige categories 3.1, 3.2, and 7.1).

It is interesting to note that the themes running through the four perspectives of the Balanced Scorecard as adapted to the SAM are similar to those of the GPRA and Executive Order 13327: improved efficiency, cost effectiveness, customer satisfaction, and results-driven performance.

ADDITIONAL PERFORMANCE INDICATORS FOR CONSIDERATION

Financial Measures

Operating cost is a key performance indicator used by private-sector organizations that is desirable for use in federal facilities asset management. Some of the considerations in developing an operating cost measure include determining what categories of costs are to be included, the resources (staff, time, funding) required to gather the data, how accounting and management systems would need to be modified to consistently gather the required data, and the costs and benefits of doing so.

Two indicators for operating costs have been developed for the SAM. The first, **Facilities Operating Current Replacement Value Index**, represents the level of funding provided for the stewardship responsibility of an organization's facilities. The indicator is expressed as a ratio of annual facility maintenance operating expenditures to CRV. The second indicator, **Facilities Operating Gross Square Feet Index**, is expressed as a ratio of annual facilities maintenance operating expenditures to the total gross square feet of the facilities portfolio. Further study is required to determine how best to calculate an indicator of operating costs that would be applicable to most agencies and cost effective to generate.

The authors also suggest development of a **Facilities Operating Gross Agency Expenses Index**. This indicator would track annual facility maintenance and operating expenditures as a function of an organization's total expenditures for the programs and people housed in the subject facilities.

A key component of facilities operating costs is utility cost. By fiscal year 2010, under Executive Order 13123, Greening the Government Through Efficient Energy Management (June 3, 1999), Federal agencies are to (1) significantly reduce energy consumption from 1990 baseline levels and (2) increase their use of renewable energies. However, for reasons already mentioned utility costs and energy consumption typically are not tracked.

The SAM offers two measures of energy costs and usage that may have value for Federal agencies. The **Energy Usage Index** is calculated as the total number of British thermal units (BTUs) consumed annually divided by gross square feet of total facilities portfolio. The BTU is an energy consumption metric that is commonly considered a worldwide standard measure; all fuels and electricity can be converted to their respective BTU content for the purpose of totaling all energy consumption. Facilities are a major consumer of energy for heating, cooling, lighting, and routine equipment operation. An energy usage index plotted over time can help to measure the energy efficiency of the portfolio and determine whether energy consumption is increasing or decreasing.

SAM's **Energy Reinvestment Index** tracks annual expenditures on energy efficiency measures as a function of the annual agency energy expenditure. It can help to measure the outcomes of investments in energy conservation efforts.

Process Improvement Measures (Internal Perspective)

To fully measure the outcomes of facilities asset management activities, indicators related to the planning, design, construction, operation, renewal, and disposal of facilities are needed.

A promising process indicator for prioritizing projects and funding to support an organization's overall mission is the **Mission Dependency Index (MDI)**. MDI uses the operational risk management techniques of probability and severity and applies them to facilities in terms of interruptability, relocateability, and replaceability. It also takes mission intradependencies (those that reside within a command) and mission interdependencies (those that reside between commands) into account. It does this through structured interviews with command representatives of individual units that cover a finite geographical area. The reported MDI is calculated using a standardized formula.

MDIs are applied at the building level, and the resulting index is a driver for prioritizing projects. Now being tested at the site level, the MDI is being designed for application at the facilities portfolio level. The MDI scores range from 0 to 100, with associated colors for visual interpretation of reports: blue (0-40), green (40-55), yellow (55-70), orange (70-85), and red (85-100, most critical). The MDI was initially developed by the Naval Facility Engineering Service Center (NFESC), and is being collaboratively refined by the Coast Guard and the Navy. GSA and NASA are also considering its use.

Workforce Support Measures (Innovation and Learning Perspective)

Understanding of the relationships between worker health, productivity, and the quality of the workplace is expanding through research. Finding ways to measure the effects of facilities investments on worker health and productivity is becoming increasingly important. Efforts are underway to develop such measures, as described below.

The **Work Environment Index**, as contemplated in the SAM, is based upon a survey instrument originally developed by IBM Consulting Group and the University of California. It assesses organizational climate within a specific work unit and to some extent includes departmental-level information relating to facilities.[1] This survey instrument covers the following areas: demographic data about respondent; communications; compensation; customer service; decision making; diversity; leadership; morale; performance management; teamwork; training and development; vision, values and business principles; and mission. Results from these surveys are tallied in the aggregate and then a high and low distribution score is developed for specific areas.

The U.S. Coast Guard is testing a **Suitability Index** that measures gaps between appropriateness of a facility to the mission requirements. The Suitability Index draws on an ASTM International[2] standard for many types of buildings that employs over 100 scales and 340 features for possible valuation. It reports a side-by-side comparison between an occupant's functionality requirements (the demand side) and the facilities' serviceability requirements (the supply side) and identifies the gap between demand and supply. When developed, the Suitability Index could potentially be applied at the portfolio level to help indicate how well the portfolio is aligned with mission and how well facilities support worker productivity.

The Coast Guard is also considering the development of a **Life Safety or Building Code Index**. The purpose of such an index would be to determine how well federal facilities comply with existing life safety measures (such as fire protection) and other building codes. Such an index would help to assess the quality of the work environment and the potential for circumstances that could be harmful to building occupants. The index would be aggregated by rolling up building-level data into the total portfolio.

Table 4.1 provides a summary of additional indicators that could be of value in measuring the outcomes of investments in federal facilities portfolios.

FINDINGS

(A) Additional performance indicators for portfolio-level management are needed to measure desired outcomes for cost effectiveness, customer satisfaction, and process efficiencies. Several promising measures are under development in federal agencies. Additional indicators could be adapted from other performance measurement systems to round out a comprehensive system of qualitative and quantitative performance indicators for Federal facilities portfolio management, over time and as resources allow.

Efforts are underway within Federal agencies to develop a Mission Dependency Index, a Facilities Suitability Index, and a Building Condition Index. The Association of Higher Education Facilities Officers-APPA, the Project Management Institute, and other organizations have developed indices for Facilities Operating Current Replacement Value; Facilities Operating Gross Square Feet; Energy Usage; Energy Reinvestment; and Work Environment; among others. In choosing additional indicators for measuring Federal facilities portfolio investment outcomes, careful consideration and study should be given to the purpose to be served, how the data to support indicators will be gathered, the resources required (staff, time, funding) to gather the data, whether existing accounting and management systems will require modification, and the costs and benefits of such modifications.

---

[1] A number of Federal agencies have developed customer satisfaction surveys for post-occupancy evaluation and other processes. It is possible that such surveys could be used in place of the survey referenced in the SAM.

[2] ASTM International was originally known as the American Society for Testing and Materials.

TABLE 4.1 Additional Indicators That Could Be Developed for Use Within Federal Agencies Over the Long Term

Category of Information	Key Performance Indicator (KPI)	Calculation	Purpose of KPI
Financial	Facilities Operating CRV Index	Annual Facility Maintenance Operating Expenditures ($)/CRV	Measures annual operating and maintenance expenditures as a function of CRV.
Financial	Facilities Operating GSF Index	Annual Facility Maintenance Operating Expenditures ($)/ Total Gross Square Feet (GSF)	Measures annual facility maintenance expenditures per gross square foot of total portfolio.
Financial	Facilities Operating GAE Index	Annual Facility Maintenance Operating Expenditures ($)/ Gross Agency Expenditures (GAE)	Measures annual facilities operating expenditures as a ratio of total organizational expenditures for the programs/missions housed in the facilities.
Financial	Energy Usage	British Thermal Units (BTUs)/Gross Square Feet	Measures total use of energy (electricity and fuels) as a ratio of total square feet of the facilities portfolio.
Financial	Energy Reinvestment Index	Annual Expenditure on Energy Efficiency Measures H 100/ Annual Institution Energy Expenditure	Measures realized benefit of energy project expenditures on overall energy costs.
Process	Mission Dependency Index	See text for process description	Assesses individual facility's relative importance to an organization's missions. The resulting index is a driver for prioritizing projects and could be used for allocation of funding across a facilities portfolio.
Workforce Support	Work Environment Index	Survey-based measure	Measures employees' satisfaction with facilities across various areas, a critical foundation for employee performance.
Workforce Support	Suitability Index	Currently under development by U.S. Coast Guard	Will measure gaps between appropriateness of facilities and mission requirements.

References

APPA (Association of Higher Education Facilities Officers-APPA). 2001. Strategic Assessment Model. 2nd ed. Alexandria, VA: APPA.

Arveson, P. 1998. Online: http://www.balancedscorecard.org/Performance Measures.

FASAB (Federal Accounting Standards Advisory Board). 1996. Accounting for Property, Plant, and Equipment. Statement of Recommended Accounting Standards, No. 6. Online: http://www.fasab.gov/pdffiles/sffas-6.pdf.

GAO (General Accounting Office). 2003. *High Risk Series. Federal Real Property.* GAO-03-122. Washington, D.C.: GAO.

NRC (National Research Council). 1990. *Committing to the Cost of Ownership: Maintenance and Repair of Public Buildings.* Washington, D.C.: National Academy Press.

NRC. 1995. *Measuring and Improving Infrastructure Performance.* Washington, D.C.: National Academy Press.

NRC. 1998. *Stewardship of Federal Facilities: A Proactive Strategy for Managing the Nation's Public Assets.* Washington, D.C.: National Academy Press.

NRC. 2004. *Investments in Federal Facilities: Asset Management Strategies for the 21st Century.* Washington, D.C.: National Academies Press.

Sanford, K., and S. McNeil. 1997. Data modeling for improved condition assessment. In *Infrastructure Condition Assessment: Art, Science, and Practice,* ed. M. Saito, pp. 287-296. New York: American Society of Civil Engineers.

Weiss, C.H. 1998. *Evaluation.* 2nd ed. Upper Saddle River, N.J.: Prentice Hall.

Appendixes

Appendix A

For Immediate Release
Office of the Press Secretary
February 4, 2004

The White House
PRESIDENT GEORGE W BUSH

EXECUTIVE ORDER FEDERAL REAL PROPERTY ASSET MANAGEMENT

By the authority vested in me as President by the Constitution and the laws of the United States of America, including section 121(a) of title 40, United States Code, and in order to promote the efficient and economical use of Federal real property resources in accordance with their value as national assets and in the best interests of the Nation, it is hereby ordered as follows:

Section 1. Policy. It is the policy of the United States to promote the efficient and economical use of America's real property assets and to assure management accountability for implementing Federal real property management reforms. Based on this policy, executive branch departments and agencies shall recognize the importance of real property resources through increased management attention, the establishment of clear goals and objectives, improved policies and levels of accountability, and other appropriate action.

Sec. 2. Definition and Scope. (a) For the purpose of this executive order, Federal real property is defined as any real property owned, leased, or otherwise managed by the Federal Government, both within and outside the United States, and improvements on Federal lands. For the purpose of this order, Federal real property shall exclude: interests in real property assets that have been disposed of for public benefit purposes pursuant to section 484 of title 40, United States Code, and are now held in private ownership; land easements or rights-of-way held by the Federal Government; public domain land (including lands withdrawn for military purposes) or land reserved or dedicated for national forest, national park, or national wildlife refuge purposes except for improvements on those lands; land held in trust or restricted fee status for individual Indians or Indian tribes; and land and interests in land that are withheld from the scope of this order by agency heads for reasons of national security, foreign policy, or public safety.

(b) This order shall not be interpreted to supersede any existing authority under law or by executive order for real property asset management, with the exception of the revocation of Executive Order 12512 of April 29, 1985, in section 8 of this order.

Sec. 3. Establishment and Responsibilities of Agency Senior Real Property Officer. (a) The heads of all executive branch departments and agencies cited in sections 901(b)(1) and (b)(2) of title 31, United States Code, and the

Secretary of Homeland Security, shall designate among their senior management officials, a Senior Real Property Officer. Such officer shall have the education, training, and experience required to administer the necessary functions of the position for the particular agency.

(b) The Senior Real Property Officer shall develop and implement an agency asset management planning process that meets the form, content, and other requirements established by the Federal Real Property Council established in section 4 of this order. The initial agency asset management plan will be submitted to the Office of Management and Budget on a date determined by the Director of the Office of Management and Budget. In developing this plan, the Senior Real Property Officer shall:

(i) identify and categorize all real property owned, leased, or otherwise managed by the agency, including, where applicable, those properties outside the United States in which the lease agreements and arrangements reflect the host country currency or involve alternative lease plans or rental agreements;

(ii) prioritize actions to be taken to improve the operational and financial management of the agency's real property inventory;

(iii) make life-cycle cost estimations associated with the prioritized actions;

(iv) identify legislative authorities that are required to address these priorities;

(v) identify and pursue goals, with appropriate deadlines, consistent with and supportive of the agency's asset management plan and measure progress against such goals;

(vi) incorporate planning and management requirements for historic property under Executive Order 13287 of March 3, 2003, and for environmental management under Executive Order 13148 of April 21, 2000; and

(vii) identify any other information and pursue any other actions necessary to the appropriate development and implementation of the agency asset management plan.

(c) The Senior Real Property Officer shall be responsible, on an ongoing basis, for monitoring the real property assets of the agency so that agency assets are managed in a manner that is:

(i) consistent with, and supportive of, the goals and objectives set forth in the agency's overall strategic plan under section 306 of title 5, United States Code;

(ii) consistent with the real property asset management principles developed by the Federal Real Property Council established in section 4 of this order; and

(iii) reflected in the agency asset management plan.

(d) The Senior Real Property Officer shall, on an annual basis, provide to the Director of the Office of Management and Budget and the Administrator of General Services:

(i) information that lists and describes real property assets under the jurisdiction, custody, or control of that agency, except for classified information; and

(ii) any other relevant information the Director of the Office of Management and Budget or the Administrator of General Services may request for inclusion in the Government-wide listing of all Federal real property assets and leased property.

(e) The designation of the Senior Real Property Officer shall be made by agencies within 30 days after the date of this order.

Sec. 4. Establishment of a Federal Real Property Council. (a) A Federal Real Property Council (Council) is established, within the Office of Management and Budget for administrative purposes, to develop guidance for, and facilitate the success of, each agency's asset management plan. The Council shall be composed exclusively of all agency Senior Real Property Officers, the Controller of the Office of Management and Budget, the Administrator of General Services, and any other full-time or permanent part-time Federal officials or employees as deemed necessary by the Chairman of the Council. The Deputy Director for Management of the Office of Management and Budget shall also be a member and shall chair the Council. The Office of Management and Budget shall provide funding and administrative support for the Council, as appropriate.

(b) The Council shall provide a venue for assisting the Senior Real Property Officers in the development and implementation of the agency asset management plans. The Council shall work with the Administrator of General Services to establish appropriate performance measures to determine the effectiveness of Federal real property management. Such performance measures shall include, but are not limited to, evaluating the costs and benefits involved with acquiring, repairing, maintaining, operating, managing, and disposing of Federal real properties at particular agencies. Specifically, the Council shall consider, as appropriate, the following performance measures:

(i) life-cycle cost estimations associated with the agency's prioritized actions;

(ii) the costs relating to the acquisition of real property assets by purchase, condemnation, exchange, lease, or otherwise;

(iii) the cost and time required to dispose of Federal real property assets and the financial recovery of the Federal investment resulting from the disposal;

(iv) the operating, maintenance, and security costs at Federal properties, including but not limited to the costs of utility services at unoccupied properties;

(v) the environmental costs associated with ownership of property, including the costs of environmental restoration and compliance activities;

(vi) changes in the amounts of vacant Federal space;

(vii) the realization of equity value in Federal real property assets;

(viii) opportunities for cooperative arrangements with the commercial real estate community; and

(ix) the enhancement of Federal agency productivity through an improved working environment.

The performance measures shall be designed to enable the heads of executive branch agencies to track progress in the achievement of Government-wide property management objectives, as well as allow for comparing the performance of executive branch agencies against industry and other public sector agencies.

(c) The Council shall serve as a clearinghouse for executive agencies for best practices in evaluating actual progress in the implementation of real property enhancements. The Council shall also work in conjunction with the President's Management Council to assist the efforts of the Senior Real Property Officials and the implementation of agency asset management plans.

(d) The Council shall be organized and hold its first meeting within 60 days of the date of this order. The Council shall hold meetings not less often than once a quarter each fiscal year.

Sec. 5. Role of the General Services Administration. (a) The Administrator of General Services shall, to the extent permitted by law and in consultation with the Federal Real Property Council, provide policy oversight and guidance for executive agencies for Federal real property management; manage selected properties for an agency at the request of that agency and with the consent of the Administrator; delegate operational responsibilities to an agency where the Administrator determines it will promote efficiency and economy, and where the receiving agency has demonstrated the ability and willingness to assume such responsibilities; and provide necessary leadership in the development and maintenance of needed property management information systems.

(b) The Administrator of General Services shall publish common performance measures and standards adopted by the Council.

(c) The Administrator of General Services, in consultation with the Federal Real Property Council, shall establish and maintain a single, comprehensive, and descriptive database of all real property under the custody and control of all executive branch agencies, except when otherwise required for reasons of national security. The Administrator shall collect from each executive branch agency such descriptive information, except for classified information, as the Administrator considers will best describe the nature, use, and extent of the real property holdings of the Federal Government.

(d) The Administrator of General Services, in consultation with the Federal Real Property Council, may establish data and other information technology (IT) standards for use by Federal agencies in developing or upgrading Federal agency real property information systems in order to facilitate reporting on a uniform basis. Those agencies with particular IT standards and systems in place and in use shall be allowed to continue with such use to the extent that they are compatible with the standards issued by the Administrator.

Sec. 6. General Provisions. (a) The Director of the Office of Management and Budget shall review, through the management and budget review processes, the efforts of departments and agencies in implementing their asset management plans and achieving the Government-wide property management policies established pursuant to this order.

(b) The Office of Management and Budget and the General Services Administration shall, in consultation with the landholding agencies, develop legislative initiatives that seek to improve Federal real property management through the adoption of appropriate industry management techniques and the establishment of managerial accountability for implementing effective and efficient real property management practices.

(c) Nothing in this order shall be construed to impair or otherwise affect the authority of the Director of the Office of Management and Budget with respect to budget, administrative, or legislative proposals.

(d) Nothing in this order shall be construed to affect real property for the use of the President, Vice President, or, for protective purposes, the United States Secret Service.

Sec. 7. Public Lands. In order to ensure that Federally owned lands, other than the real property covered by this order, are managed in the most effective and economic manner, the Departments of Agriculture and the Interior shall take such steps as are appropriate to improve their management of public lands and National Forest System lands and shall develop appropriate legislative proposals necessary to facilitate that result.

Sec. 8. Executive Order 12512 of April 29, 1985, is hereby revoked.

Sec. 9. Judicial Review. This order is intended only to improve the internal management of the executive branch and is not intended to, and does not, create any right or benefit, substantive or procedural, enforceable at law or in equity, against the United States, its departments, agencies, or other entities, its officers or employees, or any other person.

GEORGE W. BUSH

THE WHITE HOUSE,

February 4, 2004.

Appendix B

Biographies of Consultants

John H. Cable, R.A., PMP, is an architect with over 35 years of experience in managing projects of various size and complexity. His activities have been both domestic and international and have included management consulting; management of his own planning, design, and construction business; formulating and managing building energy conservation research; managing a large-scale congressionally mandated program in energy conservation standards; and teaching.

He is currently executive director of the Graduate Project Management program at the Clark School of Engineering, University of Maryland. John initiated the graduate program in project management during the fall of 1999 and an undergraduate Citation in Project Management in 2002. He teaches courses in project management fundamentals, and managing projects in a dynamic environment. John is also vice chair of the Project Management Institute's Global Accreditation Center Board of Directors and a member of the science council of NASA's Center for Project Management Research. This past summer John was an invited member of GSA's Project Management Working Group, which formulated recommendations on establishing a project management framework for the federal government.

Prior to joining the University, John was a research fellow in the Logistics Management Institute's (LMI) facilities and engineering management group, where he managed a variety of lead assignments analyzing facility design and construction practices, conducted benchmarking and business process reengineering studies, assessed the use of information technology in the management of design and construction, managed business and program planning assignments, and trained and assisted clients in becoming certified in compliance with ISO 9000 quality management standards.

Prior to LMI, John created and managed a successful design-build firm specializing in renovation and new construction of residential, commercial, and retail properties. In 1980 he was cited by *Engineering News-Record* as "one who served in the best interests of the building industry" for his work in creating the building systems energy conservation research program at the U.S. Department of Energy. In 1992 he was selected by *Remodeling Magazine* as one of the 50 best remodeling contractors in the United States.

John is a graduate of Clemson University and Catholic University and is a doctorial candidate in project management at the University of Maryland. He is a licensed architect and is certified by the Project Management Institute as a Project Management Professional (PMP). John is currently managing research on project performances metrics and is doing research on effective project performance reporting techniques.

Jocelyn S. Davis is an experienced nonprofit executive, having served both as the chief financial officer of the ICMA Retirement Trust and Retirement Corporation and as the chief financial officer of AARP, formerly the American Association of Retired Persons. In addition to the traditional responsibilities of the chief financial officer role, she has substantive operational experience, including high-volume cash transaction processing systems, call center operations, and development of specialty teams to define and resolve major business issues in a variety of contexts.

These projects have included implementation of daily valuation, consistent with the mutual fund industry, for a nonprofit investment advisor and pension administrator; rehabilitation of a call center that handles calls from 33 million association members; acquisition of a headquarters building; development of a unique investment product for a nonprofit that allowed further investment diversification; and design and implementation of Web-based credit card processing for a large national nonprofit. Her experience includes identifying and recruiting the right people for the right job at the right time and establishing an effective, collaborative team to get the job done.

Jocelyn serves currently as an independent trustee for the Allmerica Investment Trust, as an independent member of the Investment Committee of the American Psychological Association, and as a member of the board's executive committee and chairman of the audit committee of the Vaccine Fund. The Vaccine Fund works with the Global Alliance for Vaccines and Immunization to fund basic vaccines and immunizations for the poorest countries.

In addition to ongoing management consulting and board work, Jocelyn is trained as a coach and uses the principles of positive psychology championed by Dr. Martin Seligman of the University of Pennsylvania. She is currently continuing with her coaching training with MentorCoach. She offers a unique blend of practical executive-level work experience to all her coaching clients.

Jocelyn is a graduate of the College of William and Mary and has practiced for many years as a certified public accountant with Ernst & Whinney, now Ernst & Young. She works actively with clients to assist them in board governance matters from structure through director selection and evaluation to implementation of Sarbanes-Oxley in nonprofit organizations.

She is a member of the American Society of Association Executives, the National Association of Corporate Directors, and BoardSource. She is a founding member of the Advisory Board of the NPO (Non-Profit Organization) Cooperation Circle in Washington, D.C., which provides a monthly forum for peer-to-peer knowledge networking within the nonprofit community.

Appendix C

Engineered Management Systems and BCI

Engineered Management Systems

The U.S. Army's Engineer Research and Development Center, Construction Engineering Research Laboratory (ERDC-CERL) has developed Engineered Management Systems (EMS), a family of tools that aid in assessing the condition of facilities and allocating funding. Each EMS provides a functional manager with automated procedures and tools to support the planning, programming, and budgeting of infrastructure facilities maintenance and repair. The fundamental idea of an Engineered Management System is sustainment, including timely repair or replacement of system components, to meet the design or expected service life, thereby avoiding untimely and excessive recapitalization costs.

The Navy has used EMS PAVER for many years to manage all naval airfield pavements, and the system is being put in place to manage roads and parking lots. The EMS ROOFER and RAILER are in use at some activities to manage roofs and railroad tracks.

EMS BUILDER and WHARFER, which will be applied to buildings and waterfront structures, respectively, are in development and beta application testing. EMS UTILITIER is planned for eventual development. All EMSs provide a procedure for calculation and use of condition indexes to assess the condition of the facility components and systems.

The EMS condition index is mathematically determined from identified distress types germane to the component section being assessed. Distresses are objectively defined flaws in the component section. Deduct values are amounts deducted from the theoretical component condition index of 100 for a given type, quantity, and size of distress. Deduct values for each EMS distress type have been determined through research.

Building Condition Index (BCI)

In BUILDER the CI for buildings is developed from a roll up of system component sections[1] to component, to system, and finally to the building's Condition Index (BCI). The CI provides an equivalent granularity to the traditional FCI in that the roll-up BCI is applicable to a single facility or structure.

---

[1] The component section is the management unit of the EMS structure. Component sections are system elements that share common material, age, or condition. Component sections become logical maintenance management units.

APPENDIX C

The BCI is an objective rating of the condition of the component, system, or facility and is not influenced by the personal bias of the individual inspector who specifies the scope of a corrective action. The disciplined nature of the distress survey requires the inspector to identify observed distresses from a finite list of well-defined possibilities. The inspector records a severity and density range for each observed distress from the appropriate list of predefined possibilities. The EMS determines the condition index from the sum of the modified deduct values associated with the observed distress. As a result, EMS condition indices are consistent and repeatable with little variation among trained inspectors and are not influenced by subjectively determined corrective action costs.

The EMS BCI also provides the needed metric to assist in maintenance planning. Quality maintenance planning requires knowing where and when to make a sustainment, restoration, and modernization (SRM) expenditure and how much money should be allocated. EMS directly calculates a BCI and then estimates costs for SRM expenditures as a function of component replacement or repair cost and BCI. Building components are inventoried and cost models are assembled from industry standard sources. Based on research-derived algorithms using BCI, estimated costs for component-section repair or replacement are calculated according to work rules as a percentage of component-section replacement value or predetermined unit costs.[2] EMS uses a family of cost calculations that are specific to facility system, component, or component-section. Using these costs, BUILDER identifies a list of work items. Each work item is associated with an EMS-calculated component-section BCI. Each inventoried component-section that is below a user-defined standard condition is a candidate for corrective action.

---

[2]Both PAVER and RAILER set up work policies for correcting distresses or defects. Unit costs for these activities are derived from job order contracts, indefinite delivery contracts, or other sources. ROOFER also addresses distress fixes.

Appendix D

Space Utilization Index

The Space Utilization Index (SUI), under development by the U.S. Coast Guard, compares actual to allowable space. Actual space measurements are made during a space utilization assessment. Allowable space is based on the commandant space standards, which define approved space allowances on the basis of personnel and other factors. The SUI is calculated by dividing actual square feet by allowable square feet.

The SUI measures compliance with commandant space standards, which ensures equitable distribution of space (and funding) across the agency. The SUI can be used to collect data on existing space and to update the standards to better support operations and mission readiness objectives. An SUI of 1.00 means the space exactly complies with commandant space standards. In practice an SUI range is established to account for reasonable departures for the standards due to suitability issues and local variations. For example, an SUI range between 0.95 and 1.15 is considered reasonable.

In practice SUI will be used along geographical boundaries, command structures, and space categories. Each perspective will provide insight into how the space is being used, identify gaps or opportunities, and provide a structure to pursue targeted space planning and management activities.